ASME Press Robotics Engineering Book Series

Designs and Prototypes of Mobile Robots

Edited by
Marco Ceccarelli
Emin Faruk Kececi

ASME PRESS

ASME Press books are available at special quantity discounts to use as premiums or for use in corporate training programs. For more information, contact Special Sales at CustomerCare@asme.org

A catalog record is available from the Library of Congress.

Print ISBN: 978-0-7918-6047-2
ASME Order No. 860472

Series Editors' Preface

Dear Reader,

As you are likely aware, robotics has become one of the key areas of research and engineering in the decades. It is already significantly integrated into numerous aspects of human everyday life and will continue to profoundly influence the quality of life in the 21st Century by pioneering new paradigms in science and implementing them in fundamentally novel robotics engineering technologies. Thus, it is evermore necessary to proficiently train incoming generations of engineers who would be capable to address future challenges and move robotic science and engineering forward. In this regard, ASME Press launched the Robotics Engineering Book Series and thereby established a unique avenue to facilitate technological discoveries by producing top quality technical books and texts.

I am delighted to introduce the first two books in the Series: *Designs and Prototypes of Mobile Robots* and *Mobile Robots for Dynamic Environments*. Authored by a group of well-known researchers led by Prof. Marco Ceccarelli and Dr. Emin Faruk Kececi, these books go beyond the typical robotics application and uniquely integrates mechanical design and control decisions within interactive dynamics of the operator-machine-environment system. I believe that the historical overviews, recent developments and technical approaches presented in these books will inspire the next wave of novel thoughts and ideas to be further researched and engineered.

Another distinguished feature of these books is its team of researchers, composed of veterans in robotics engineering as well as prominent professionals of younger generations. The Robotics Engineering Book Series is committed to both assisting eminent scholars in publishing their findings and introducing outstanding new researchers to the world of books in Robotics. New titles are coming soon, and we look forward to close collaboration with researchers and engineers around the world.

Vladimir V. Vantsevich, ScD, PhD, ASME Fellow
Founding Editor, ASME Press Robotics Engineering Book Series

January 20, 2015

ASME Press Robotics Engineering Book Series

Advisory Group Members

Contributing Authors

Editors

Marco Ceccarelli
LARM: Laboratory of Robotics and Mechatronics, University of Cassino and South Latium, Italy

Emin Faruk Kececi
Istanbul Technical University, Department of Mechanical Engineering; Turkey

Authors

Qiang Huang, and Si Zhang
Intelligent Robotics Institute, Beijing Institute of Technology, China

Xilun Ding
Robotics Institute, Beijing University of Aeronautics and Astronautics, China

Fan Yang
Robotics Institute, Beijing University of Aeronautics and Astronautics; China

I-Ming Chen, Wei-Ting Yang, Albert Causo, Song Huat Yeo
School of Mechanical and Aerospace Engineering, Nanyang Technological University, Singapore

Jorge Solis
Department of Engineering and Physics, Karlstad University, Karlstad, Sweden
Research Institute for Science and Engineering, Waseda University, Tokyo, Japan
Humanoid Robotics Institute, Waseda University, Tokyo, Japan

Contents

Preface

For several decades now, mobile robots have been integral to the development of new robotic systems for new applications, even in non-technical areas. Mobile robots have already been developed for such uses as industrial automation, medical care, space exploration, demining operations, surveillance, entertainment, museum guides and many other industrial and non-industrial applications. In some cases these products are readily available on the market. A considerable amount of literature is also available; not all of which pertains to technical issues, as listed in the chapters of this book.

Mobile robots will always be further developed with the goal of performing locomotion tasks, those related to movement and interaction with the surrounding environment, within which a task can be fulfilled even without the supervision of human operators. The complexity of locomotion requires different solutions both for design and operation. As such, a large variety of mobile robots and mobile robotic systems has been, and still can be, developed. In fact, considerable advancements have been achieved within the last few decades, and a vast amount of literature is already available detailing a large variety of mobile robots. The literature emphasizes design issues, operational success, procedures and algorithms that can be used specifically for these applications, as opposed to general approaches for a variety of cases.

One key point for mobile robots is interaction with the environment in which the mobile robot moves and corresponding solutions can determine the success or failure of the motion. Indeed, the mechanical design is not very often considered a critical issue, but rather it is often included as an issue in the overall design of mechanical solutions within servo-controlled operation and environment interaction. A second important issue is the acceptance of robotic systems and the corresponding psychological aspects, when robots are proposed to operators and users in fields with very low levels of technical means in their current work practice.

These two subjects are the core of the discussions in this book and its companion volume, *Mobile Robots for Dynamic Environments* (available separately from ASME Press), which aims to illustrate not only the potential but also the problems for the dissemination of mobile robots and mobile robotic systems in all human activities with service aims.

Authors have been invited from all over the world and chapters have been selected after review as to approach the most challenging aspects and applications of mobile robotic systems, with the aim to survey the current state-of-the-art and its future potential.

We believe that readers will enjoy this book and its companion, and will utilize the knowledge gained with satisfaction and will be assisted by its content in their interdisciplinary work for engineering developments of mobile robots, in both old and new applications. This book and its companion can be used as a graduate level course books or guide books for the practicing engineer who is working on a specific problem which is described in one of the chapters.

We are grateful to the authors of the chapters for their valuable contributions and for preparing their manuscripts on time. Also acknowledged is the professional assistance by the staff of ASME Press and especially by Dr. Vladimir Vantsevich, who has enthusiastically supported this book project, as the Robotics Series Editor.

We are also thankful to our families for their patience and understanding, without which the realization of this book and its companion, with the input of so many experts from different fields and countries, might not be possible. Moreover, our students hard and dedicated work taught us a lot.

Dr. Marco Ceccarelli and Dr. E. Faruk Kececi
Cassino, Italy and Istanbul, Turkey - July 2014

1. Linkages for leg mechanisms

Marco Ceccarelli, University of Cassino and South Latium, Italy

Abstract: Walking robots are based on the mechanical structures of legs, in which mechanisms play an important role in functionality and performance characteristics. These robots have been developed with a wide variety of solutions over time. A short account of the history of using mechanisms in walking robots is presented in this paper, illustrated through significant examples and an explanation of primary concepts, with the goal of stressing challenges for future development.

1.1 Walking issues for robots

Locomotion in nature is ensured with legged systems, as demonstrated by animals with a wide range of solutions and performance characteristics. In general, animals have symmetric structures featuring two, four, six or eight legs that ensure stable walking and even running within a large variety of environments [1]. Figure 1-1 presents a summary of the animals that have generally been considered an inspiration for artificial walking machines since such applications began to be developed.

In addition, a three-legged operation is sometimes used, as illustrated by someone walking with the help of a cane or a kangaroo using its tail for slow motions [2]. Figure 1-2 shows examples of those natural structures and operations in three-legged systems.

Although the number and anatomic structure of the legs can be very different among different types of animals, kinematic design models and operational functions can be summarized as in the following: having a foot with a space mobility of at least 3 d.o.f.'s and an ovoid-like trajectory of ankle reference point, as shown in Figure 1-3. In particular, the kinematic design of a human-like leg can be characterized by a spherical joint S at the hip articulation and revolute joints R at knee and ankle articulations. The walking operation can be characterized by looking at the articulation motions as per the joint angles, and by evaluating the leg movement as per the trajectory of an ankle reference point. The walking movement path can be summarized in a typical trajectory of the ankle reference point that shows an ovoid-shaped path whose dimensions give

a)

b)

c)

d)

Figure 1-1 Animals that are considered inspiration in designing and operating walking machines: a) ostrich; b) horse; c) crab; d) spider.

the step size S and step height H, as indicated in Figure 1-3(b). The step size gives the capability of the walking motion, while the step height H gives indication of overpassing obstacles while walking. Both parameters can be increased on purpose during operation, but H is the main characteristic of a walking leg that makes legged systems very flexible, even for use in environments crowded with obstacles.

The end-effector of a leg is a foot, which has the function of contact and interaction with the ground in order to provide a proper action and reaction in walking, both for force exchange and stable positioning. The peculiarities of foot anatomy and operation are also significant in differentiating and specializing animals for the specific environments in which they usually live.

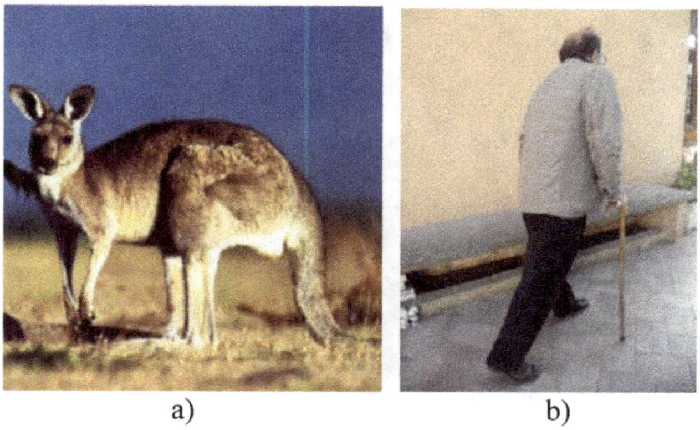

Figure 1-2 Three-leg structure for walking in nature: a) a kangaroo in slow motion; b) a person using a cane to walk.

The anatomy of legs in nature can be summarized as a bone structure with joint articulations that are activated by a complex system of muscles whose operation is regulated by a complex neurological system, as illustrated in Figure 1-4(a), referring to human anatomy.

However, from a kinematic viewpoint leg functionality can be summarized as due to the bone structure with its articulations, which can be understood as revolute joints in the ankle and knee, and as a spherical joint in the hip, as illustrated in Figure 1-3(a).

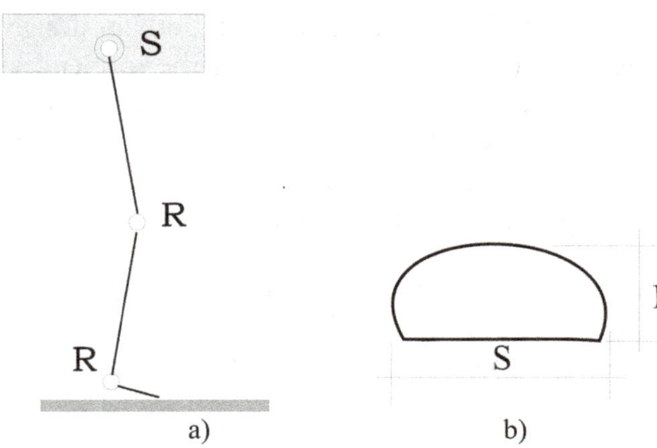

Figure 1-3 Main characteristics of a kinematic design of a human-like leg: a) structure; b) reference ankle point path.

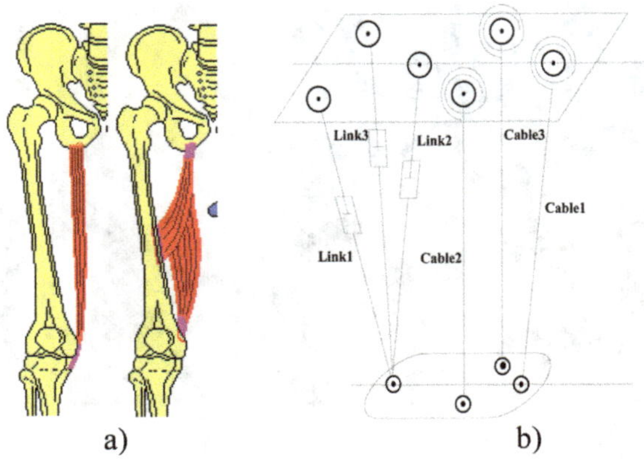

Figure 1-4 Human leg structure: a) sketches of human anatomy; b) a kinematic model considering bones and muscles.

Muscle function can be modelled as action by links or cables that pull the bone at which they are attached, and more complex leg models can be identified as structures that are based on parallel manipulator architectures [3], as illustrated in Figure 1-4(b).

In replicating leg anatomy and/or functionality in walking machines and robots, bone structure is often used as a reference and kinematic designs are developed to achieve walking locomotion with motion properties for smooth operation, payload capability, and nature-like actions. Thus, attention is focused on replicating a suitable human-like path for a reference foot point.

Problems for leg mechanisms can be outlined, both for design and operation purposes, as linked to each other, by tackling the problem as in mechanics of robots [4], and they can refer to demanding situations for overpassing obstacles or climbing steps. Mechanical aspects can be identified, as indicated in the sketches in Figure 1-5, as referring to possible solutions for mobile robots with legged solutions in attaching the following problems:

- motion synchronization for step size and lift height
- balancing actions and dynamics simulation
- ground-foot contact with consideration of friction and impact aspects
- actuation and forward velocity
- motion planning (also for obstacle avoidance)
- sensored and controlled interactions with the task of walking

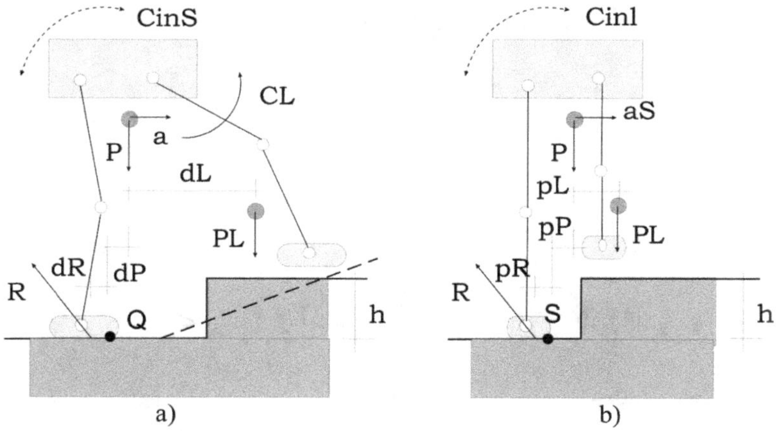

Figure 1-5 A scheme for leg design in overpassing an obstacle: a) sagittal plane; b) front view.

- design compactness and light weight
- payload capacity and environment interaction

Overcoming obstacle conditions as a leg design problem can be discussed by referring to the schemes in Figure 1-5, when referring to climbing a step as or for obstacle overpass.

The instantaneous static equilibrium gives the necessary conditions for walking and overcoming obstacles of the maximum allowed step height. Referring to Figure 1-5(a), in a sagittal plane the equilibrium can be expressed by:

$$
\begin{aligned}
&R_h - P\frac{a}{g} - PL\frac{a_L}{g} \geq 0 \\
&R_v - P - PL \geq 0 \qquad (1) \\
&R_v dR - PdP - PL(dP + dL) - C_{inS} \geq 0
\end{aligned}
$$

where R_h and R_v are the horizontal and vertical components of R contact reaction; P is robot weight; PL is leg weight; a is motion acceleration and g is the gravity; dL, dR, and dP are the indicated distances; CL is the torque actuating a leg; C_{inS} is the sagittal component of the inertial torque due to waist balancing movement. Point Q is assumed as the foot contact point about which the system will rotate in a possible fall. In the front plane, as illustrated in Figure 1-5(b), the equilibrium can be expressed by:

$$R_l - P\frac{aS}{g} \geq 0$$

$$R_v - P\frac{a_S}{g} \geq 0 \qquad (2)$$

$$R_v pR - PpP - PL(pP + pL) - C_{inl} \geq 0$$

where R_l is the lateral component of R; a_S is the lateral acceleration of robot body; C_{inl} is the lateral component of the inertial torque of waist balancing movement. Point S is assumed as the foot contact point about which the system will rotate in a possible fall. The components R_h and R_l refer to friction actions at the foot contact area. By using Equations (1) and (2) expressions can be deduced for design and operation features that are useful to overcome obstacles of height h. From geometric viewpoints the obstacle/step of height h can be overpassed when the leg mobility gives:

$$l_1(1 - \cos\phi_1) + l_2(1 - \cos\phi_2) > h \qquad (3)$$

in which l_1 and l_2 are the lengths of leg links, whose angles ϕ_1 and ϕ_2 are measured with respect to a vertical line. Thus, in general the design problem related to overpassing obstacles can be formulated by using (1) to (3) to properly size the leg links, and to give required mobility ranges and actions for proper leg operation, as any other robotic mechanisms [4].

Specific analysis, also for design purposes, and specific aspects for legged systems are treated in the next chapters of this book. In this chapter mechanism structures for leg designs are outlined using an illustrative approach to the historical evolution, in order to stress the variety and role of the mechanical designs of mechanisms in walking machines, and also challenges for the future.

1.2 A historical survey of mechanisms for walking machines

Mankind has always dreamed of having an easy, no-effort means of transportation, and since ancient times devices have been conceived and built for transportation of passengers and materials, often with the use of animals. In addition to load transportation (including carrying passengers), there have been challenges for locomotion purposes, with attempts being made to replicate and/or mimic solutions in nature, as seen with a large variety of animals. Transportation systems were developed mainly as wheeled systems that are today very successful in vehicle technology, with different structural designs and operational issues, as compared with walking machines. Thus, since the evolution of wheeled systems demonstrates very different characteristics from those of walking machines, in this chapter the attention will be focused specifically on walking machines. Readers can refer to corresponding literature for the history of vehicles, like cars, trucks, motorcycles and so on.

With the development of Robotics, locomotion systems and walking machines have received stronger and stronger attention, with a large number of inventions and designs mainly within the last three decades. But most of the concepts and even structural designs have evolved, even unconsciously, from past solutions. Historical studies have been published on the evolution of walking machines in general, also within the history of robotics, whose literature is reported with basic references, for example in [5–7].

In this chapter notes are outlined for a history of walking robots by looking specifically at the kinematic design of the mechanical structure. The goal is to stress a significant role of mechanism design in the development of walking machines, even with modern robotic features. The significant role of mechanism design of leg structure is evident as related to motion purposes in walking. Mechanisms are fundamental to perform such a motion capability with suitable performance, and their designs can achieve suitable solutions also in terms of efficiency and mimicking natural actions.

Since the subject is of considerable extension both from timeline viewpoints and variety of examples, this chapter has been organized with few significant examples to stress main points of a historical evolution by using an illustration approach.

Walking machines have always been developed with a strong inspiration from nature, and the historical developments can be outlined mainly by looking at application purposes and the machine technology that was used in their designs.

As regarding the applications, walking machines historically have been designed and used mainly for entertainment, and only marginally as a specialized means of transportation. Indeed the two applications could even coincide, as in the case of automata for theatre plays or amusements for the upper classes, including governors and kings. But most of the time practical operations were quite limited, both in motion range and payload capability. This was due in part to the available technology, mainly for a power source of the input motion. In general, automata were developed as based on simple mechanism (like slider-crank mechanisms), with gears and walking capabilities quite limited. Later, during the Renaissance and even more so in the following centuries automata reached a significant complexity and even versatility.

Spring systems and weight combinations were used for a long time to store energy to be used in walking machines for autonomous operation. Indeed a strong advance in walking machines and their possible uses occurred when motors became available with compact design but considerable power. This began with steam engines in the 19th century, continued with electric motors, and the current solution is servo-controlled actuators. The availability of proper power sources for on-board solutions promoted the use of more complicated mechanical designs, with high number of link mechanisms in order to give more efficiency, payload capability, and flexibility.

In addition to the above aspects of actuation and application, walking machines attracted interest with successful designs because their structures were based on clever mechanism solutions. Even today the mechanism structure plays an important role, both in the functionality and efficiency of walking machines with robotic features.

In the following emblematic examples are reported and briefly discussed to show the main characters of the technological historical evolution of walking machines within the above-mentioned general aspects.

While most of the automatic/robotic devices from ancient times are lost, they can be found in documents, artistic representations and even in humanistic literature. An example from the time of classic Greece is represented in Figure 1-6(a), based on studies in [8]. The stand-up motion for the automaton is not obtained by a direct leg system, but rather a mechanism produces the motion for a human-like action, as sketched in Figure 1-6(b). The mechanical design has been reconstructed as in Figure 1-6(a) with gears and simple linkage, utilizing the mechanism-based technology available at the time. Indeed most of the automata with a leg-based motion, as they were built along many centuries, have been designed with mechanical structures, including geared systems and simple linkage transmissions. The human-like operation is

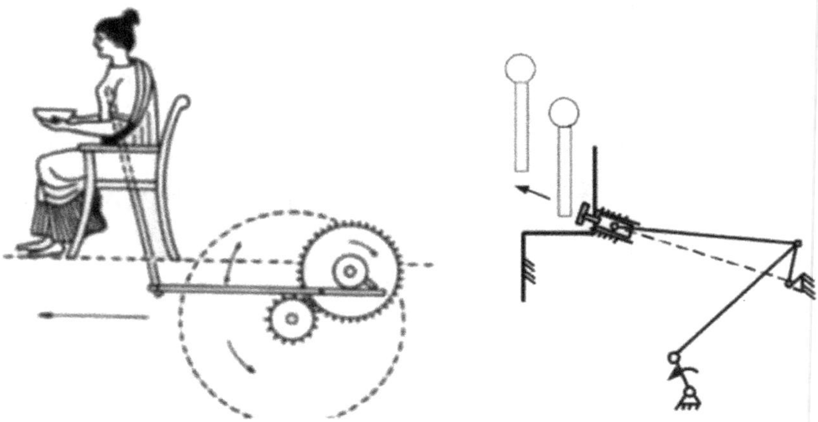

Figure 1-6 The Automaton Nysa designed in Alexandria in the 3rd Century B. C. Source: [8] (with permission); a) a modern reconstruction; b) a kinematic study of mechanism functioning.

usually obtained for the overall figure and not specifically for a linkage-based mechanism structure.

Automata with mechanical designs were enhanced beginning in the time of the Renaissance. This was as an indication of advanced skills in developing fairly automatic systems for practical purposes, in leisure and entertainment, and even in religion. Emblematic of is the example of the lion mechanism that Leonardo Da Vinci designed and built to greet the king of France, with a leg design shown in Figure 1-7 [9]. In this case the leg structure is limited and the advanced functionality is demanded to the transmission mechanism within the walking machine.

Figure 1-8 [10], is an example of the many designs that were developed in 18th century as automata, mainly for leisure and entertainment purposes. Main of them were installed in luxury clock designs in order to show the flow of time. The mechanism structure is based on linkages for the leg motion whereas gears and springs are used for storing and transmitting the necessary mechanical energy. Linkages for legs are of fairly simple structures, while the gear transmissions can be composed of complex trains both for proper motion laws and compact designs.

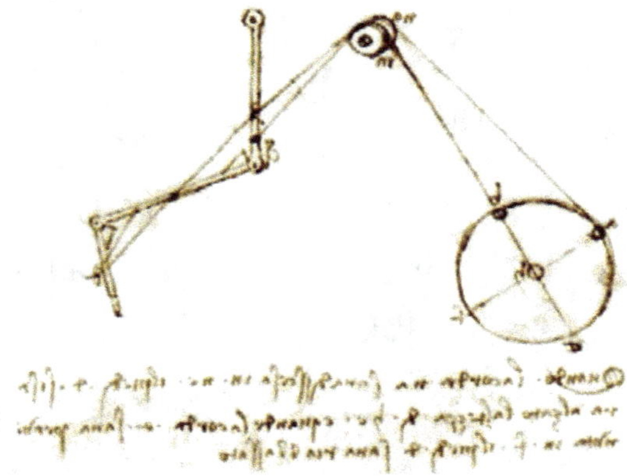

Figure 1-7 A drawing of leg design for lion automaton by Leonardo Da Vinci at the beginning of the 17th century.

The automata designs for legged machines experienced renewed interests in the second half of the 19th century when smaller steam engines were attempted. Even the mechanism technology that was the core of Industrial Revolution evolved, and more complex mechanisms were

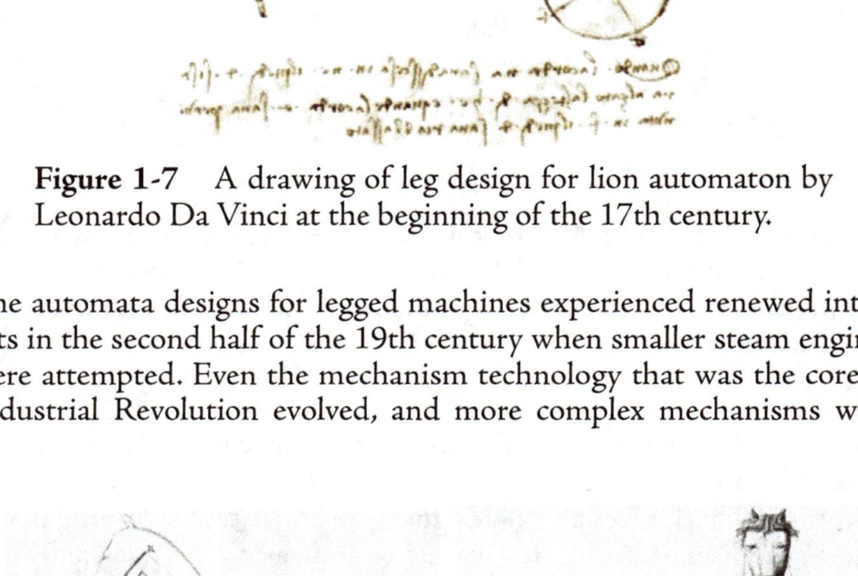

Figure 1-8 Gear-based design of a mechanical horse in an automaton of the 18th century, as reported in Borgnis (1818–21) [10].

conceived and applied in walking machines, as seen in the design patents in Figure 1-9.

Emblematic is the kinematic solution by Chebyshev at the end of the 19th century that shows a design which is still the inspiration of new mechanisms to this day [11]. It is a linkage design with 1 dof whose characters are used even in what today is known as kinetic art, where mechanisms move within a sculpture or are the sculpture itself. In fact, linkage mechanisms (like the one in Figure 1-10) were conceived up to the middle of the 20th century in several solutions, as reported in [11, 12], and are still the core of the mechanism designs of walking machine today. In particular, pantographs are often used with the important role of amplification of an input motion with suitable trajectory features for the leg end-point, as indicated for example in [13, 14].

But technological evolution was not only important in the western world. For example, in China the culture also reached heights that required mature technology. In the field of automation Chinese designers

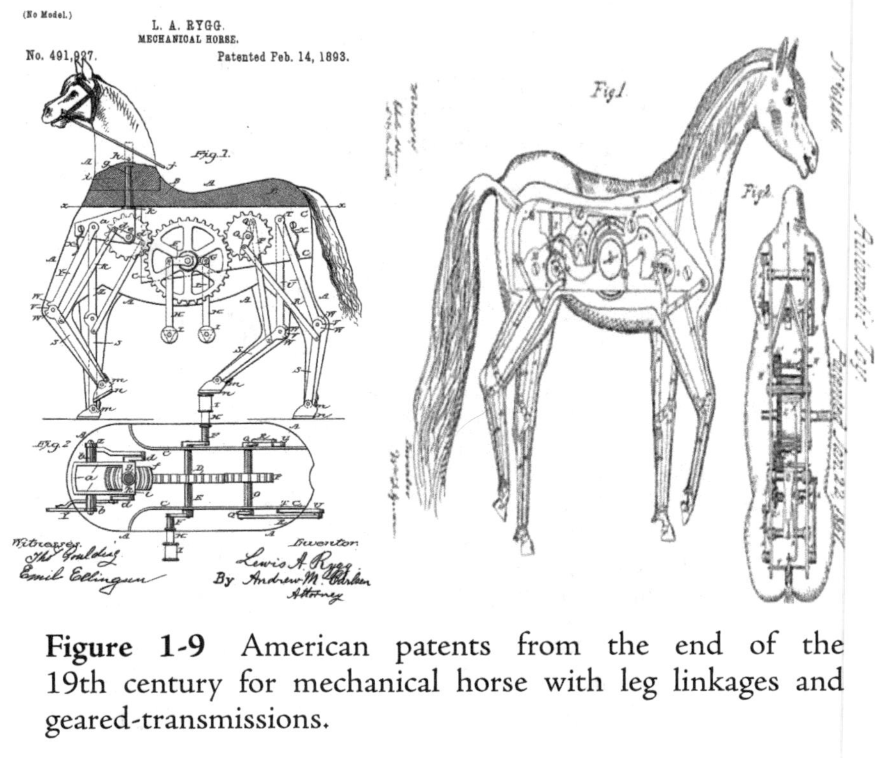

Figure 1-9 American patents from the end of the 19th century for mechanical horse with leg linkages and geared-transmissions.

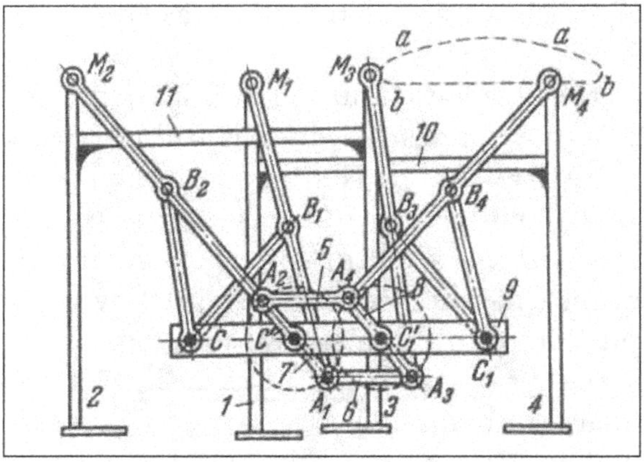

Figure 1-10 Linkage design for 1 dof walking machine by Chebyshev at the end of the 19th century [11].

developed brilliant solutions; a very significant example is the Wood Cow pictured Figure 1-11. It was used for the transportation of heavy loads by using 1 dof walking machine.

In Japan, since the 18th century Karakuri designs were developed for religious exhibitions and leisure applications with complex mechanisms, including partially walking motion performance. Leg design was limited to structures whose motion was activated by brilliant solutions in compact mechanical designs.

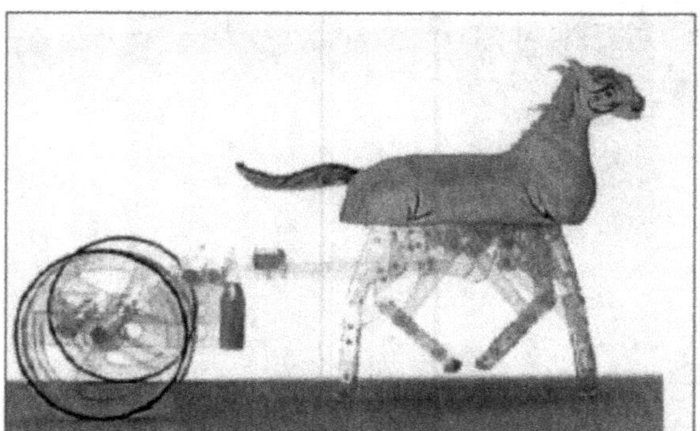

Figure 1-11 A reconstruction of the Chinese Wood Cow built in the 5th century BC (from Ancient Chinese Machines Foundation at National Cheng Kung University in Tainan).

1.3 Modern solutions

Since the main goal of a walking machine is locomotion that is based on leg motion, in modern times specific fields have been identified as being related to walking machines and humanoid robots. The specific interests have also produced specific forums of discussion and a resulting dissemination of information, both in conferences and publications, such as journals and books.

In particular, humanoid robots were started in the 1970s, with pioneering designs developed by Ichiro Kato at Waseda University in Tokyo, Japan [15], as seen in Figure 1-12. Since then the leadership in this area has been centered in Japan. The anthropomorphic design of legs was obtained with simple open kinematic chain structures that are actuated with complex actuation systems for proper motion control.

Today the more sophisticated humanoid robots, like those in Figure 1-13, still have such a kinematic structure in the legs but with more powerful efficient actuation for advanced sensor-controlled motion (also due to more compact powerful robot-directed actuators).

a) b)

Figure 1-12 Early biped solutions for humanoid robots at Waseda University from the 1970's to 80's. Source: [15] (with permission); a) pneumatically actuated, b) servo-controlled fluidically actuated with inverse pendulum stabilizer.

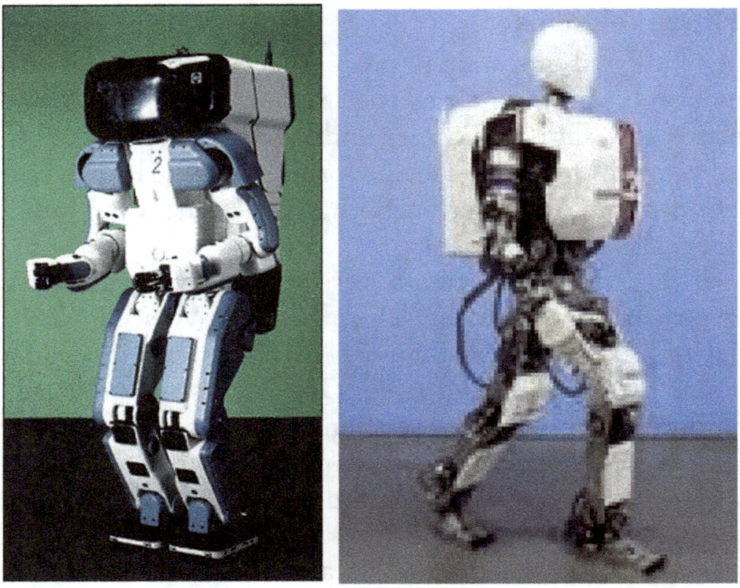

a) b)

Figure 1-13 Humanoid robots. Source: [1] (with permission); a) Honda robot built in 1993; b) Waseda robot Wabian-2 in 1997 version.

Linkage mechanisms are still used in multi-leg walking machines, like those in Figure 1-14, replicating the animal solutions in Figure 1-1. In fact, all kind of linkages are used in leg designs, with 1, 2 and 3 dofs solutions, in walking machines with 2, 4, 6 and 8 legs. These have been used for different applications, ranging from entertainment purposes up to exploration, including monitoring scopes, as discussed in the next chapters.

Emblematic are the walking machines for humanoid robots in Figure 1-15. They were developed with two different solutions for high payload transportation purposes for human users. In Figure 1-15(a) a linkage mechanism is based on a pantograph and parallelograms to achieve both high payload capability and 1 dof actuation for each leg [16]. In Figure 1-15(b) a parallel manipulator architecture is implemented with high payload capability and high motion versatility in replicating the human leg structure of bones and muscles [15], as in the sketch in Figure 1-4(b).

The above few illustrated examples can be considered representative of a large production of designs and prototypes of walking machines

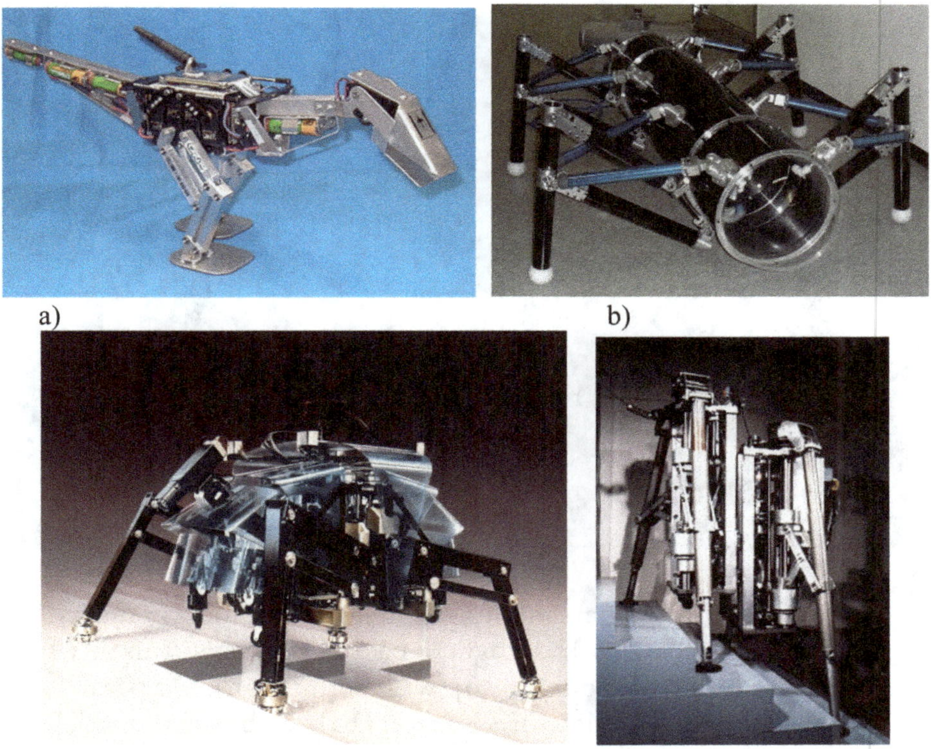

a) b) c) d)

Figure 1-14 Walking machines with pantograph-based leg mechanisms as designed by Shigeo Hirose. Source: [17] (with permission); a) a biped dinosaurian robot, [18]; b) a hexapod robot; c) a spider-like four-leg TITAN IV robot, d) compact quadruped TITAN VI robot.

and humanoid robots that have been experienced successfully as used mainly within the last three decades. The mechanism structures of the legs have evolved from pure mechanical solutions to sophisticated sensored servo-controlled systems, but still with a mechanical design as the core for locomotion functionality.

1.4 Challenges for future developments

Since the main goal of a walking machine is recognized in the locomotion as based on leg functioning, it is well understood that future improvements also strongly depend on the efficiency of the locomotion action and leg design.

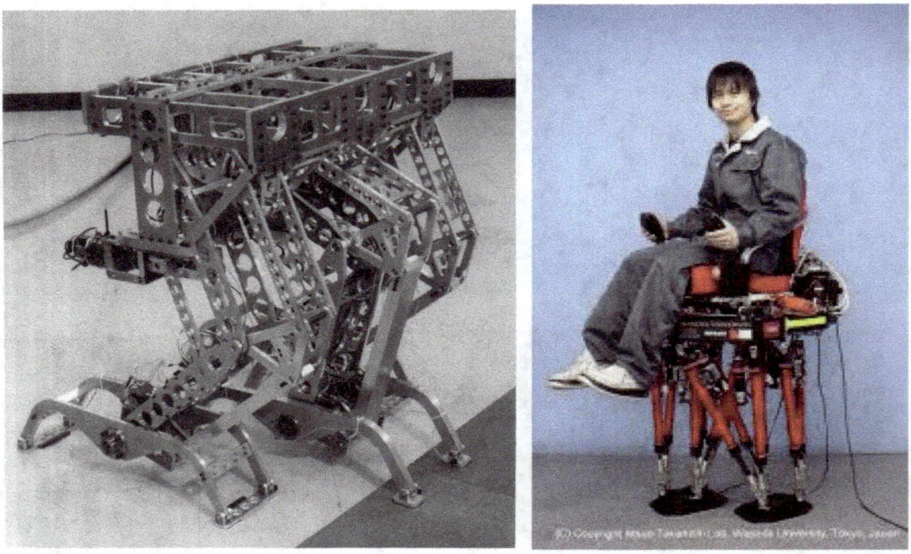

a) b)

Figure 1-15 Mechanism designs in novel locomotion units in humanoid robots: a) pantograph based leg mechanisms in a three-leg solutions. Source: [16] (with permission); b) a parallel manipulators architecture. Source: [15] (with permission).

Locomotion action is performed by combining mechanical functioning with operation of hardware and software systems that power, support, regulate, and monitor the locomotion. Similarly a leg design is obtained in efficient solutions when its mechanical design is successfully supported and integrated by the above mentioned hardware and software systems. Those systems make a leg design a mechatronic system, since they are related to aspects and requirements that in an integrated way power, support, regulate, and monitor the action of a leg. Discussion on those systems, and challenges for their improvements in enhanced solutions of walking machines, are presented in the next chapters of this book, which also refer to specific mobile robot categories.

Challenges are currently under consideration for improving the compactness of the nature-based solutions by looking at a power increase, both in terms of payload capability and walking frequency within compact mechanical designs. Biomechanics' study of locomotion in nature can be still of great help, and indeed great attention is still focused on looking for biomimetic solutions.

Nevertheless, as in the past, future developments will depend strongly on the available technology for single components and overall system integration. Trends and interests are already well-defined as focused on smart materials, sensing equipment and power units that can determine walking machines, even to the point of being competitive with wheeled machines and vehicles. Mechanism structures are also under reconsideration not only for better solutions but for suitable integration with the expected advances in other components.

Figure 1-16 summarizes the concepts that future developments of walking machines depend on, advances in several aspects and components with integrated considerations and implementations where the

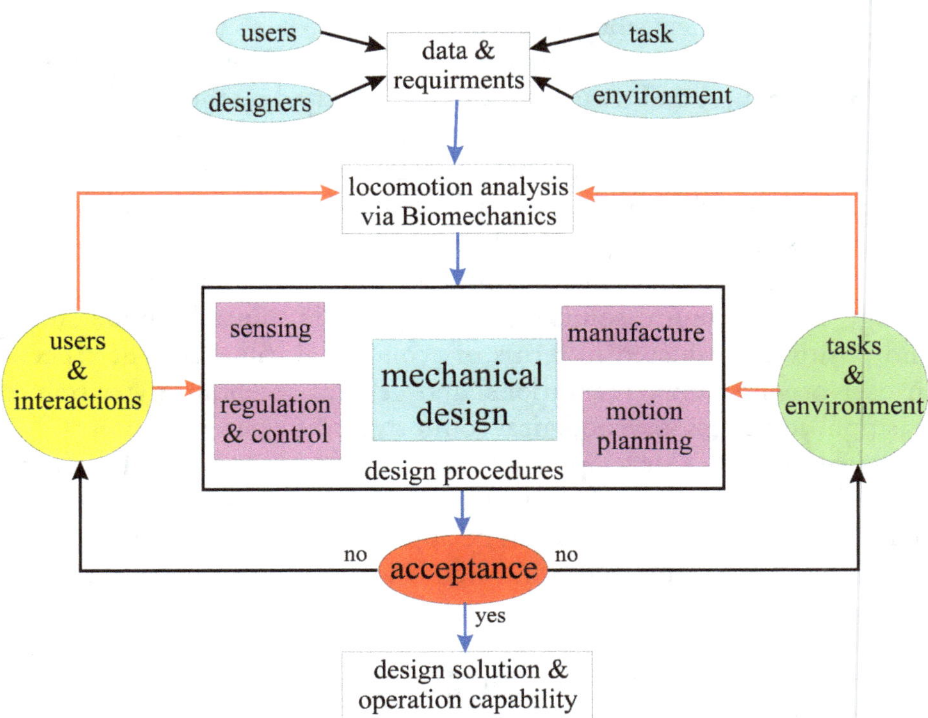

Figure 1-16 A general scheme for design of leg systems with the central role of mechanical structures.

mechanical structure will still pay a central role, both in design issues and operational efficiency.

Data and requirements will be increasingly linked to prescriptions, expectations and constraints that can be defined by general considerations, but also by goals for end-users and designers who take into account the specific task and environmental impact of an application.

Locomotion analysis will still look for biomimetic inspiration since it is well recognized as the optimization in nature solutions. Biomechanics still plays an important role in defining actions and motions that an artificial locomotion system must perform, not only for stable efficient walking but even for achieving compact light designs similar to natural systems.

The mechanical design requires design procedures that are also oriented to user- friendly operations, with multidisciplinary aspects integrating problem solutions with mechatronic features, not only for structural design but also for manufacturing issues, both with market components and new smart materials.

Considerations by end-users can again be an important influence in design procedures and their solutions, as indicated by the left circle box in the diagram of Figure 1-16, where emphasis is stressed on aspects that are related to human-machine interactions. Similarly, the right circle box indicates the influence of task data and environment conditions in all the aspects of design procedures.

Finally, the acceptance question is a crucial point in modern technology since technical solutions, although valid, must be understood and positively evaluated in terms of economical value, benefit aspects in user-oriented implementations, and general acceptance viewpoints (fashion, aesthetics, psychological comfort, etc.) by potential users.

However, future developments and the widespread use of walking machines may depend to a large degree on the enhancement of leg design in the many aspects that can be approached with new perspectives and technologies. Challenges and trends for designing the operation of leg mechanisms will be focused on improvements to:

- compact design
- efficiency
- payload capability
- flexibility
- environment impact

and they can be outlined mainly in the following topics:

- topology of mechanism structures
- formulation for high-speed computation
- light compact mechanical design
- interactions with environments and human users or operators
- advanced sensorization
- adaptable control systems
- smart materials and components
- anthropomorphic and nature-inspired solutions

The above considerations have emphasized expected advances in traditional performance, but perhaps new applications will be invented for walking machines as a function of the new technology that will be made available in the future. In addition, future developments will generate the rise of new problems and new emerging topics from other fields. A final goal to be considered is achieving not only nature-like designs but the development beyond standard solutions with different solutions.

1.5 Conclusions

This chapter presented a short account of the history of walking machines in terms of leg mechanisms, using an illustrative approach in order to outline their main characteristics over a long time and the evolution of technology. It is relevant to note that interest arose in different cultures and over different time periods with the same aim, developing systems imitating the nature solutions in animals for applications that only recently can be recognized as having significant practical implementations. The role of mechanism structures has been and is still prominent, since mechanisms are the core of walking machines in performing the biomechanical operation of locomotion. Therefore, challenges and trends for more capable walking machines, even with full robotic designs, seem to be focused on better and more efficient reproduction of locomotion mechanics, beside a system integration with technological advances for the other components of walking machines.

1.6 References

[1] Morecki A., Waldron K.J., *Human and Machine Locomotion*, Springer, New York, 1997.

[2] Liang C., Ceccarelli M., Carbone G., Novel Biologically Inspired Tripod Walking Robot, *13th WSEAS International Conference on Computers*, Rhodos, 2009, paper no. 620–141, pp. 83–91.

[3] Ceccarelli M., Carbone G., A Study of Feasibility for a Leg Design with Parallel Mechanism Architecture, *IEEE/ASME Conference on Advanced Intelligent Mechatronics*, Singapore, 2009, paper, no. 131.

[4] Ceccarelli M., *Fundamentals of Mechanics of Robotic Manipulation*, Kluwer Publ., Dordrecth, 2004.

[5] Rosheim, M.E., *Robot evolution*, Wiley, New York, 1994.

[6] Ceccarelli, M., A Historical Perspective of Robotics Toward the Future, *Fuji International Journal of Robotics and Mechatronics*, 2001, Vol. 13, No. 3, pp. 299–313.

[7] Bautista Paz E., Ceccarelli M., Echavarri Otero J., Munoz Sanz, J.J., *A brief illustrated history of machines and mechanisms, Science and Engineering*, Book series on History of Machines and Machine Science, Vol.10, Springer, Dordrecht, 2010.

[8] Koetsier T., Kerle H., The Automaton Nysa: Mechanism Design in Alexandria in the 3d Century B. C., *CD Proceedings of 2013 IFToMM Workshop on History of MMS*, Palermo, 2013, paper no. 08.

[9] Taglialagamba S., *Leonardo da Vinci: Automation and Robotics*, CB Editions, Poggio a Caiano, 2010.

[10] Borgnis G.A., *Traitè complet de mecanique appliqée aux Arts*, Bachelier, Paris, 1818–1821, 9 Volumes.

[11] Artobolevsky I.I., *Mechanisms in Modern Engineering*, Mir Publ., Moscow, 5 Vols., 1975–80.

[12] Chironis F., *Mechanisms, Linkages, and Mechanical Controls*, McGraw-Hill, New York, 1961.

[13] Ryan A.D., Hunt K.H., Adjustable Straight-Line Linkages – Possible Legged Vehicle Applications, *Journal f Mech., Transm., and Automation in Design*, July 1985, Vol. 107, pp. 256–261.

[14] Song S.M., Waldron K.J., *Machines that Walk: The Adaptive Suspension Vehicle*, MIT Press, Cambridge, 1988.

[15] Hashimoto S., Takanishi A., (Eds), *Humanoid Robotics Institute at Waseda University*, Waseda University Publishing Office, Tokyo, 2008. (Web page of the Humanoid Robotics Institute at Waseda University, http://www.humanoid.waseda.ac.jp/, Tokyo, 2014.)

[16] Funabashi H., Takeda Y., Kawabuchi I., and Higuchi M., Development of a walking chair with a self-attitude-adjusting mechanism for stable walking on uneven terrain, *Proc. Tenth World Congress on the Theory of Machines and Mechanisms*, Oulu, 1999, pp. 1164–1169.

[17] Hirose S., Robotics Research in Hirose-Yoneda Laboratory, Tokyo Institute of Technology, Tokyo, 2001.

[18] Takita K., Katayama T., Hirose S., Development of Dinosaur-like Robot TITRUS -Its Dynamics and the Motion Utilizing the Dynamic Effect of the Neck and Tail-, *IEEE/RSJ International Conference on Intelligent Robots and Systems*, Las Vegas, 2003, pp. 607–612.

2. Exoskeletons and bipeds

Qiang Huang and Zhangguo Yu, Beijing Institute of Technology, China

Abstract: Robotics is a field of interdisciplinary research, and the development of its relevant disciplines has promoted its rapid progress in recent decades. Exoskeletons and bipeds, which are inspired by humans, have attracted many researchers. Exoskeletons, as the extenders of humans, can enhance a person's strength, power and endurance. They are usually utilized in the medical field to provide assistance for disabled people. Bipeds are capable of assisting humans to do repeatable tasks or to operate tasks in dangerous environments. The significance of studying these two kinds of robots is not only to break through the limitations of human capability, but also to gain a better understanding of human body. This chapter focuses on the history and key technologies of exoskeletons and bipeds. Some representative robots and mechanisms are described in detail.

2.1 Exoskeletons

Exoskeleton robots are external structural mechanisms whose joints correspond to those of the human body. They are worn by an operator, and the physical contact between the operator and the exoskeleton allows direct transfer of mechanical power and information signals between them [1]. Generally, exoskeletons are divided into lower extremity exoskeletons [2], upper extremity exoskeletons [3] and full-body powered exoskeletons [4].

Compared to other robots, exoskeleton robots can make full use of both the capability of human judgment and the capability of robotic motion. Specifically, the human neural network is one of the most efficient and quickly responding systems, it can accomplish many complicated tasks, and it works well in many fuzzy conditions [1]. Robotic manipulators can exert large forces and torques due to the structure and power of their actuators, but they lack flexibility. Thus, the outer mechanical garment would dramatically amplify the operator's strength and endurance [5], and significantly enhance its adaptation in the unstructured environment, compared with autonomous robotic systems [6].

Meanwhile, although biologists have discovered the elaborate structure of human body, the ability to understand and imitate human movements is still at a primitive level. Building exoskeletons is a good way to understand these queries, and the study of exoskeletons helps with the design of prosthetics. Thus, we can gain new insights into those problems and learn more about possible solutions.

2.1.1 History and overview

In the 1960s, in cooperation with the United States military, General Electric Co. (GE) developed the first exoskeleton robot, Hardiman, as seen in Figure 2-1(a), which was intended to be used in a practical environment [7]. It was driven by hydraulics and electricity, and supposed to amplify the operator's strength 25 times. The control system of the exoskeleton was a kind of slave-master system. Unfortunately, the project was not successful because of the limitations of safety and reliability issues, and the exoskeleton was never tested on humans [8].

Made in 1969 at the Mihailo Pupin Institute, Powered Exoskeleton, seen in Figure 2-1(b), was the first exoskeleton robot that could walk actively. It was pneumatically powered and partially cinematically programmed to generate human-like gait.

Figure 2-1(c) shows a successful version of an active exoskeleton robot, realized and tested at Belgrade Orthopedical Clinic in 1972, for

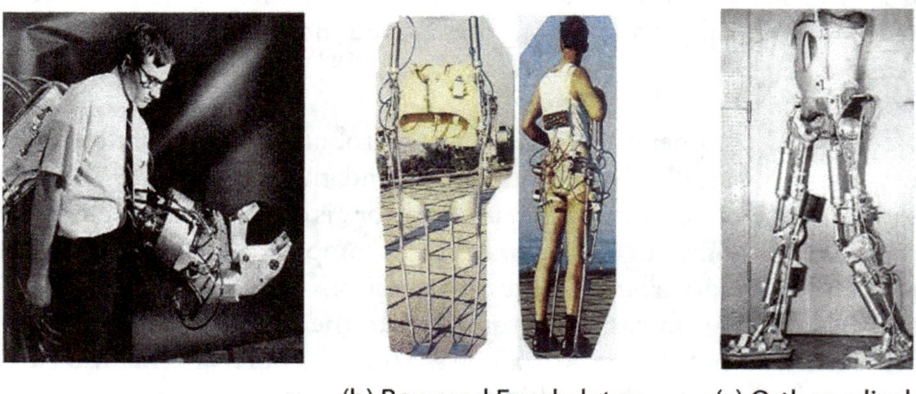

(a) Hardiman. [Courtesyof GE]

(b) Powered Exoskeleton. [Courtesy of Mihailo Pupin Institute]

(c) Orthopedical. [Courtesy of Belgrade Orthopedical Clinic]

Figure 2-1 The early era of the exoskeleton.

the rehabilitation of paraplegics and similarly disabled persons. It was pneumatically powered and electronically programmed.

In past decades, some well-known and innovative exoskeletons were developed. The HAL (Figure 2-2), developed by the University of Tsukuba [9], was an exoskeleton robot whose purpose was both performance-augmenting and rehabilitation. It was designed to be integrated with humans and utilized to assist people with gait disorder. The EMG sensors on a human's leg muscles were adopted to control HAL's electric actuators (EMG signals of human muscles were studied to help understand the motion intention of the operator [10]). The anthropomorphically-based BLEEX (Figure 2-3) focused on the lower

Figure 2-2 The HAL. [Courtesy of University of Tsukuba.]

Figure 2-3 The BLEEX. [Courtesy of Berkeley.]

extremity exoskeleton [11]. The applications of BLEEX were providing soldiers, disaster relief workers, wildfire fighters, and other emergency personnel the ability to carry heavy loads. BLEEX was the first load-carrying lower extremity exoskeleton in the world.

Miguel A.L. Nicolelis and his laboratory in Duke University are developing technologies that allow electrical signals from brain to control robotic limbs [12]. If they succeed with this project, a paralyzed teenager may wear the exoskeleton to make the first kick of 2014 World Cup (Figure 2-4). The main idea behind this kind of exoskeleton is that paralyzed people might control their limbs just by thinking.

2.1.2 Mechanism design principles

The motion of exoskeleton robots is motivated by the actions of humans, and one does not have to be concerned with collisions between human and exoskeleton. Thus, the basic rule in selecting the structure of an exoskeleton robot is to match its degrees of freedom (DOFs) with that of a human. Considering this rule, the best way to design an exoskeleton robot is to build it with exactly the same DOFs as a human. However, there are more than 300 DOFs in the human body, and so the prospect of building and controlling such a complicated machine seems impossible. With the two purposes mentioned above, the function of exoskeleton robots is to support the motion of legs and arms. And the strength-assists of exoskeleton needn't support every single muscle of

Figure 2-4 The first robot used by a paralyzed teenager. [Courtesy of Duke University.]

the human body, but rather the entire movement of a human. These facts are greatly helpful in simplifying the DOF of exoskeletons.

Exoskeleton robots are classified into two types according to different structure, anthropomorphic and non-anthropomorphic. In the anthropomorphic architecture, the structure attempts to exactly match the kinematical DOF of human [13]. Thus, when building an anthropomorphic lower extremity exoskeleton, it is better for the structure of the exoskeleton's leg to directly duplicate the structure of a human's leg. However, one major difficulty is that the joints in human legs cannot be fully duplicated in designing the trajectories of joints [14]. For instance, the human knee does not exhibit a pure rotation. Duplicating all its kinematics will result in a complicated (and perhaps non-robust) mechanical system [15]. Therefore, only the major motion in human joints is considered, except coupling the position of exoskeleton and human.

Non-anthropomorphic architectures are not widely used in exoskeletons because the structure may interfere or limit the operator. But non-anthropomorphic architectures are highly successful in other areas such as bicycles, and they open up a wide range of possibilities for structural design.

In designing an exoskeleton robot, anthropomorphic and non-anthropomorphic are not separated rigidly, since parts of the exoskeleton are anthropomorphic and some are not. This is called pseudo-anthropomorphic [16]. In the next section, a pseudo-anthropomorphic exoskeleton, which is designed for lower-extremity and upper-limb power support, is introduced.

All in all, the basic rule in designing the structure of the exoskeleton is safety. The exoskeleton must not force the operator to a position the operator cannot reach. The working range of an exoskeleton should be restricted so as to be compliable with human operators.

The BLEEX (Figure 2-5) is used as an example to explain the mechanical designing schemes. This robot focuses solely on lower extremity exoskeletons. It consists of hip, knee and ankle joints like a human, but the details of these joints are different from a human. There are seven distinct DOFs in each leg: three at the hip, one at the knee and three at the ankle.

Although the joints of humans are not purely rotational, the combination of rotational joints is able to imitate the motion of human legs. From Figure 2-6 we can see that the hip joint of an exoskeleton is a combination of three rotational joints: alternate rotation, flexion/extension

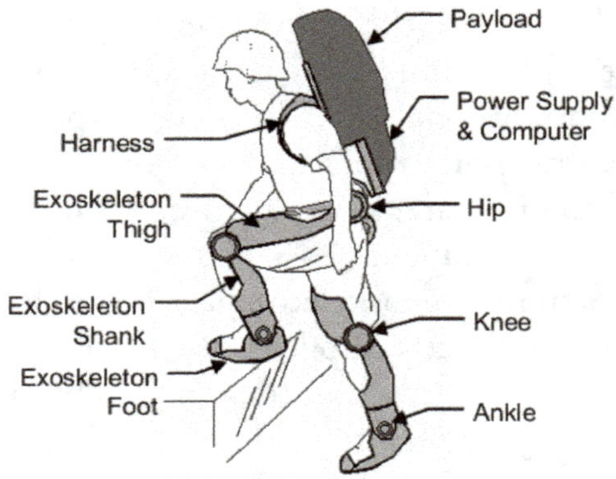

Figure 2-5 Overall structure of BLEEX. [Courtesy of Berkeley.]

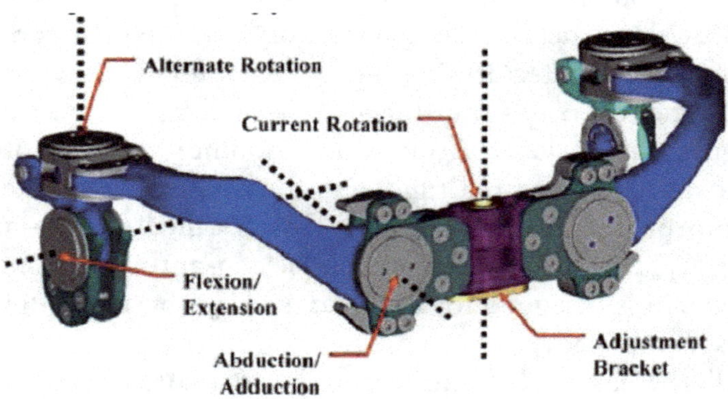

Figure 2-6 BLEEX hip degrees of freedom. [Courtesy of Berkeley.]

and abduction/adduction. This combination's movement is as a human's ball and socket joint with three DOFs. The design of the adjustment bracket allows for the adjusting exoskeleton to accommodate operators of various widths.

Figure 2-7 shows the configuration of an ankle joint. The flexion/extension axis coincides with the human ankle flexion/extension axis. This configuration makes the exoskeleton's foot compliant to flex with the human's foot.

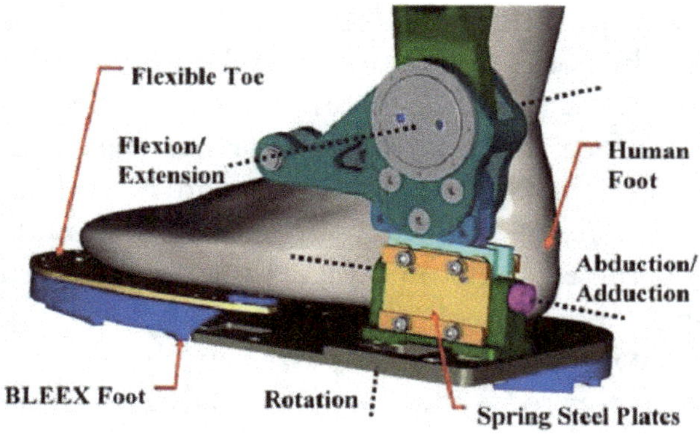

Figure 2-7 BLEEX ankle degrees of freedom. [Courtesy of Berkeley.]

2.1.3 Sensing and control algorithm

The main goal of exoskeletons is to integrate human and robot into a whole system which combines the intelligence of the human and the strength of the robot. Thus, the exoskeleton system should have the ability of quickly detecting the state of the operator and predicting its intention.

Most control methods aim to enhance the operator's power, which is one of the main functions of exoskeletons. When the intention of the operator is predicted, the power augments control module transfers the operator's intention into the movement of the exoskeleton's actuators.

To put it simply, the power augment control can be illustrated as follows. The control input is the intention, such as the force or biological signal, of the operator. The output is the joint speed or force of the system. Generally speaking, the output is about a dozen times greater than the input, which means that power augmentation is realized by a large movement obtained with small operator torque.

- Sensing
  The difficulty of directly measuring the intention of the operator makes it a tough task to estimate the operator's attention. Consequently, various sensors are utilized to detect the state of the operator for the sake of human intention prediction.

The states of the operator should include the contact force between the operator and the exoskeleton, the operator's muscle movement, the angle velocity of each joint and the body posture. All this information is essential for the exoskeleton to follow and predict the operator's motion and intention. Meanwhile, the conditions of the outer environment should also be collected, as the exoskeleton replaces the human in contacting the environment.

When concerned with low-extremity, the balance control is an essential factor to be considered, and force between the foot and ground must be measured very precisely. Thus, many kinds of force sensors, such as pressure sensors, tactile sensors and six-axis force/torque sensors, are installed on the exoskeleton to detect the contact force. Also, a displacement sensor is used to measure terminal movements of the links or end-effector of the exoskeleton.

- Impedance control

In a master-slave system [17], the interaction between master and slave can be simplified with mechanical impedance with inertia, viscosity and stiffness [18]. This means that human and exoskeleton interact with each other through a quadratic system, as the exoskeleton system is a typically master-slave model. In addition, human and exoskeleton can be dealt with the same equivalent simplification. But human operators adjust their impedance characteristics according to a given task and change control properties during operation [19]. In addition, kinematics compatibility is required between human and robot [20].

The planar model (Figure 2-8) depicts the fundamental interaction between human and exoskeleton.

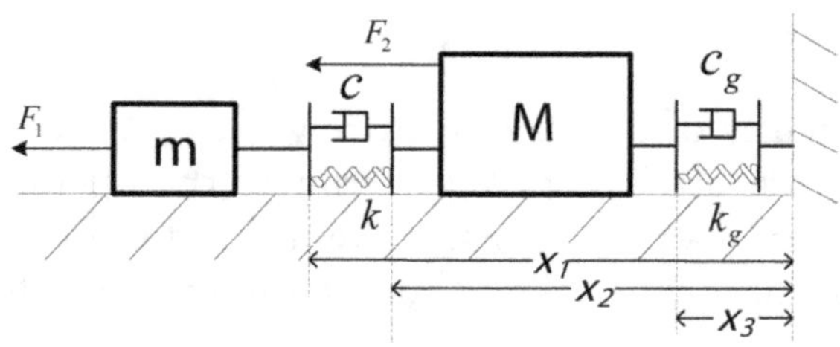

Figure 2-8 Impedance model of interaction.

According to the Newton's law,

$$F_1 = m\ddot{x}_3 + k(x_2 - x_1) + c(\dot{x}_2 - \dot{x}_1) \tag{1}$$

$$F_2 = m\ddot{x}_3 + c_g\dot{x}_3 + k_g x_3 + k(x_1 - x_2) + c(\dot{x}_1 - \dot{x}_2) \tag{2}$$

where F_1 denotes the force generated by operator muscle, F_2 denotes the force generated by exoskeleton actuator, m denotes the inertia of interaction, k denotes the stiffness of interaction, c denotes the viscosity of interaction, M denotes the inertia of load, c_g denotes the viscosity of load, and k_g denotes the stiffness of load. Note that the original lengths of both springs is zero.

If the controller and the mechanical impedance characteristics satisfy

$$k = 0 \tag{3}$$

$$\dot{x}_1 = \dot{x}_2 = \dot{x}_3 \tag{4}$$

Then the equation can be simplified as

$$F_1 = m\dot{x}_1 \tag{5}$$

$$F_2 = m\ddot{x}_3 + c_g\dot{x}_3 + k_g x_3 \tag{6}$$

Equation (5), (6) is the ideal control of a master-slave system. From the equations we can see that the slave and master part are separated from each other. With two separated parts, the operator will feel lifting the weight M as if it is m instead.

The impedance control algorithms are complicated and need some simplification in the model for computation. For example, it is hard to find a system in reality with an impedance model of zero stiffness. So to remedy this defect, the controllers of the impedance model are sometimes complicated and require a long time for computation; this may uncouple operator and exoskeleton.

Steve Jacobsen and the engineers at Sarcos propose some new ideas to control the robot XOS. Their ideas are based on a simple force control strategy to ensure the exoskeleton follows the operator with minimal interaction forces [21]. This control theory sounds quite practical. When the operator starts to move, he/she imposes forces on the exoskeleton; the controller then drives the actuators to adjust the position of joints to minimize the forces detected. So the instant operator generates a force and then the exoskeleton accomplishes the movement instead of the operator. This control algorithm is called 'get out of way' by Steve Jacobsen. It sounds easy but is hard to realize, as the value of each sensor changes several thousand times in a single second, and all the computations must be done in this short time. If not, the operator will feel like walking in the water with his/her action far behind the ideas.

The control algorithm is explained in a simplified model shown in Figure 2-9. Only two sensors are placed on the contact surface of the operator and the exoskeleton. The sensors measure the pressure between muscles and exoskeleton, and then send the signals to the controller to compute the movement of actuator to minimize the pressure of the sensors. Finally, the exoskeleton will track the trajectory given by the controller.

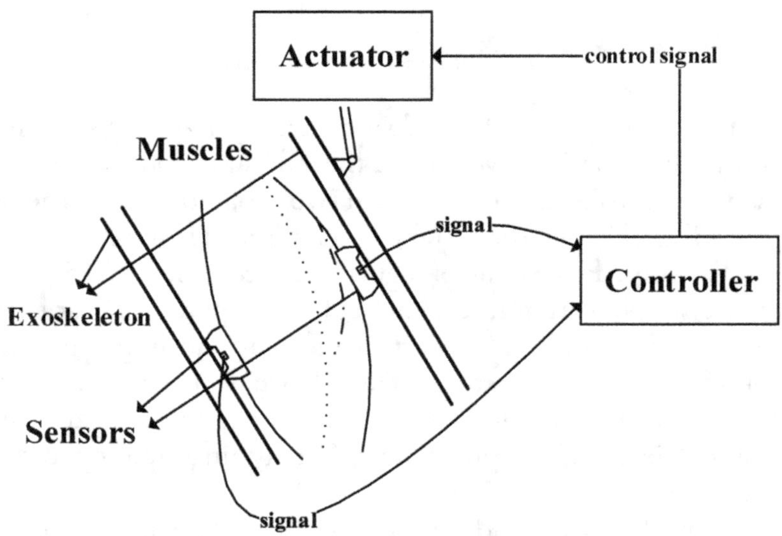

Figure 2-9 Simplified model of force control.

- Electromyography signal control
  As the force control of the exoskeleton requires quick response and is highly non-repeatable, many exoskeletons adopt biological signals, such as electromyography (EMG) signals, to measure the joint position and angular velocity. The potentiometers sensors are used to achieve this goal.
  The EMG signal (0.01–10 mV) is generated when the muscles contract and directly reflect the muscle activities of a human. The use of EMG has an advantage that the intention can be estimated before the muscle tension is exerted since the myoelectric potential rises before the muscle activation. This phenomenon is known as electro-mechanical delay [10]. So the EMG signal is treated as an effective control method for many kinds of assist robotic system, especially for the exoskeleton robots [22]. Figure 2-10 shows the fundamentals of the EMG signals [23].
- Neural interface control
  A new control algorithm is now considered with the development of sensors, computers and an understanding of the human neural system. The new control algorithms are based on the neural signal interactions between exoskeleton and human, which are quite different from the previous force control strategy. The advantage of neural interface (NI) control is that whereas force control works

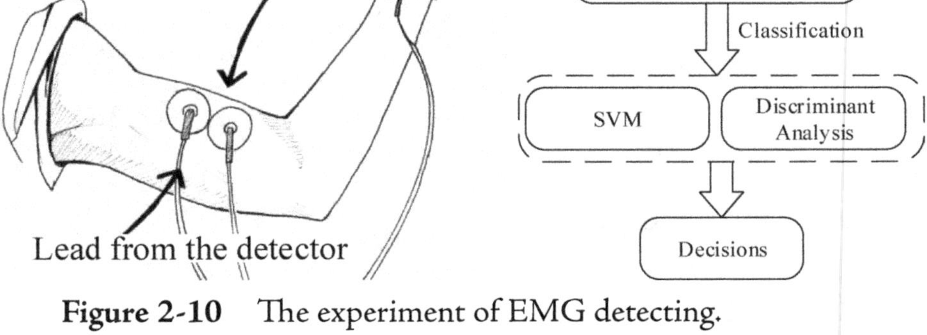

Figure 2-10 The experiment of EMG detecting.

based on movements, NI control executes when the operator has the will to execute. This means that NI control is faster. The exoskeleton and muscles receive signals from the human brain simultaneously, so that plenty of time is saved for the exoskeleton to drive its actuator and get ready for the movement. Exoskeletons with NI control also have reduced obstruction when worn.

To reduce obstruction, exoskeletons must instantaneously follow the operator's voluntary and involuntary movements. This requires the exoskeleton to couple with muscles. Figure 2-11 shows the structure of an NI control exoskeleton. The smooth operation of the whole system relies on continuous interactions between living brain tissue and artificial electronic or mechanical devices [24]. Exoskeletons act based on signals detected from the neural system, and the feedback of movement, to decrease the discrepancy of movement between operator and exoskeleton.

- Comparison among different control methods

With the increased sensitivity of the sensor and rapid response of the actuator, the impedance control theory will work well, and one will feel like the exoskeleton is a part of his/her body. There are limitations to this control method, for the performance of force control strategy is confined by the sensitivity of sensors. The increasing sensitivity of sensors will cause a loss in robustness of the control algorithm. If the sensitivity is too large, an external force or torque exerted on the exoskeleton will ruin the stability of the whole system.

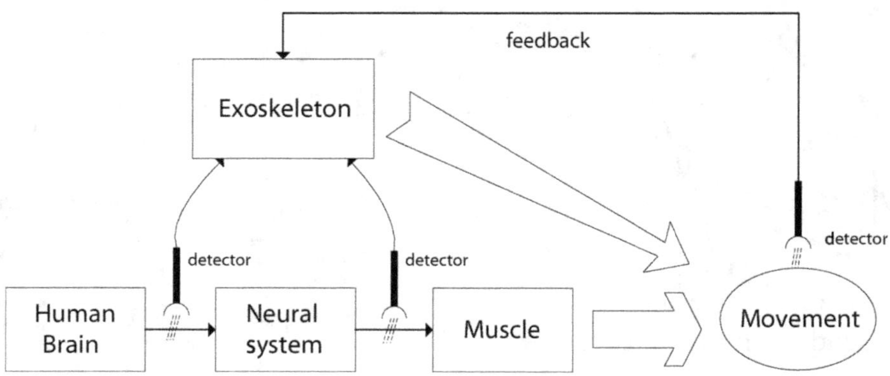

Figure 2-11 Model of NI control.

Compared with the impedance control method, the EMG control has the advantage of not requiring detection of the reaction between operator and exoskeleton. But the complex relationship between the EMG signals and torque leads to the main drawback of the EMG control method. This limitation needs an accurate and reliable model [25] that is difficult to achieve in practical applications.

NI control exoskeletons are capable of predicting the movement of humans. Exoskeletons have recorded thousands of signals from the human neural system under various conditions. Thus, when humans are accomplishing movement, exoskeletons can fore-act to the expected position to reduce obstruction. However, in practical applications, the precise detection of the signals is complicated, which means many sensors have to be placed on the user's body, and this reduces the flexibility and agility of the exoskeleton.

Above all, two factors should be noted when using these control methods. First, the relationship between human and exoskeleton is not simple rigid coupling. Actually, more compliances and additional masses in the coupling exist in reality, which complicates the dynamic behavior of the system with uncertain unstable parameters. Second, human controller parameters are difficult to estimate, and the feedback of the human is time-varying and nonlinear. Thus, vibration should be suppressed in the control system to enhance efficiency and safety.

2.1.4 Actuators and portable power supply

The efficiency and the weight of the actuators and power supply are very critical for the design and performance of the exoskeleton robot, especially for the mission duration of the portable exoskeleton. There are various actuators applied to the exoskeleton. HAL [9] uses servo motors for the hybrid assistive leg, while a hydraulic driving unit is used for the BLEEX [8]. Series elastic actuators (SEAs) and pneumatic actuators are also commonly used in the application of the exoskeletons.

- Series elastic actuators

  Series Elastic Actuators (SEAs) [26], invented by Gill Pratt and Matthew Williamson at the MIT lab, are high-fidelity force-controllable actuators, which have shown their effectiveness in many force controlled robots.

In SEAs, the output of a motor and gear train is connected through a spring. The output force of the actuator depends on the compression of the spring, which can be calculated according to Hooke's Law ($F = kx$). The output force control is then achieved by serving the compression of the spring via feedback. Hence the spring turns a position-controllable device into a force-controllable device. Figure 2-12 shows the architecture of SEA. The load sensing element in SEAs is much more compliant than a traditional load cell, which leads to much lower output impedance and higher fidelity force control.

- Electro-hydrostatic actuator

Hydraulic actuators have a relatively high control bandwidth and high specific power, as the hydraulic fluid is largely incompressible. Electro-hydrostatic actuators (EHAs) are a class of servo motor-driven displacement control type hydraulic system [27], with typical architecture shown in Figure 2-13.

Generally speaking, no servo valves are used when the pump is connected to one hydraulic actuator. A model-based approach needs to be taken to make the output power estimation for the system, which is more complicated because the coupling between components in EHAs are hydraulic and have internal leakage.

Exoskeletons with EHAs have the following properties: more sensitivity and control over force, high durability based on the large surface area transmission of force, and more actuator placement freedom. The apparent inertia of the actuator and friction

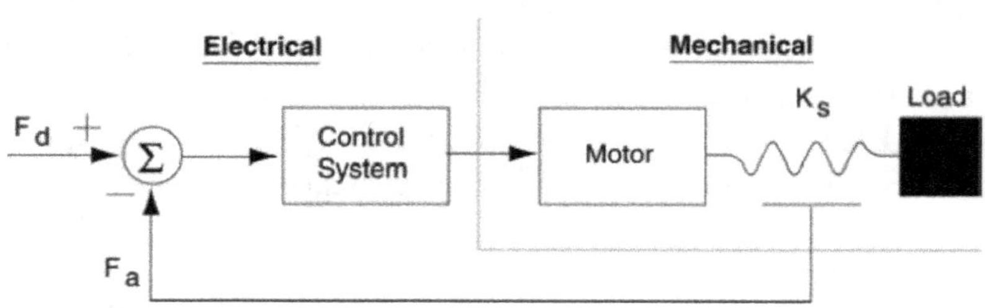

Figure 2-12 Architecture of SEA.

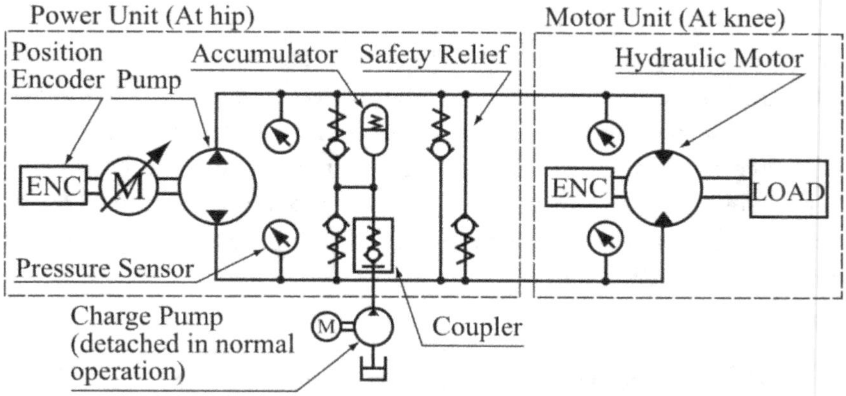

Figure 2-13 Hydraulic schematic of electro-hydrostatic actuator. [Courtesy of Nakamura et al. [27].]

are two major causes of back-drivability degradation. Reducing apparent inertia of the actuator by torque feedback, and cancelling friction torque with estimated friction torque, are effective methods for reducing these effects.

- Portable power supply
  A portable power source is not necessary when the exoskeleton is designed for therapy and not as a stand-alone, but there would be wider applications if the device could be used outside of the clinic. Up till now, a battery was still the main power source for exoskeletons and other robots. Long duration is based on both the battery system and the efficiency of the robot system, and it is a bottleneck for the mobility of the robots.

  At present, some advanced exoskeletons, such as HAL and BLEEX, have portable power supplies. However, exoskeletons with portable power will be more and more popular in the future.

2.2 Bipeds

Biped robots are robots with two legs, and the main difference between them and other robots is biped locomotion. Since the first biped robot, WABOT-1 [28], was developed nearly half a century ago, researchers have made a great deal of progress in developing these robots.

2.2.1 History and overview

The field of biped robots can be divided into two broad classes: fully actuated and passive robots. Fully actuated robots consist of two categories, according to control mode: position control and force/torque control. Passive robots are composed of passive-dynamic and limit-cycle walkers. They may be completely passive, or partially actuated, and the motions generated can be human-like and highly efficient, even with partial actuation [29, 30].

- Position control biped robots

  Position control robots, such as Honda ASIMO [31] and HRP-2 [32], are driven by motors connecting joints with reducers, and the control parameters are joint angle and angular acceleration.

  ASIMO (Figure 2-14) is an example of joint position control biped robots. At the beginning, it can complete tasks like serving tea, pushing a cart and walking on rough terrain [33]. Later, it realizes decision-making and real-time planning in uncertain circumstances, and can also run at a speed of 9 km/h. Waseda University, a pioneer in biped robots, built the world's first biped robot, WABOT-1, in 1970, and has since developed multi-generation biped robots. In 2006, the new generation biped robot WABIAN-2R (Figure 2-15) achieved human-like

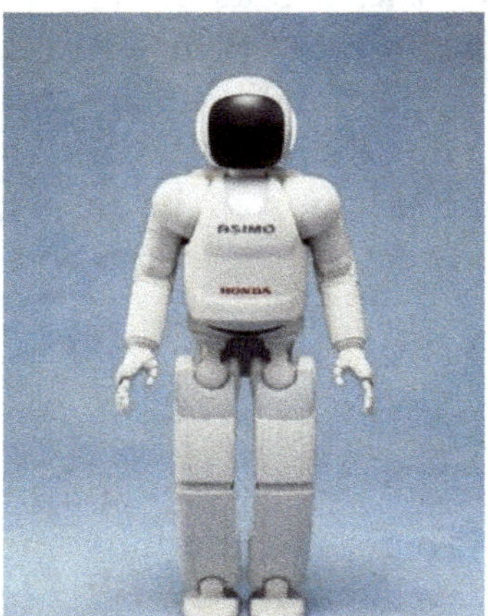

Figure 2-14 ASIMO. [Courtesy of Honda.]

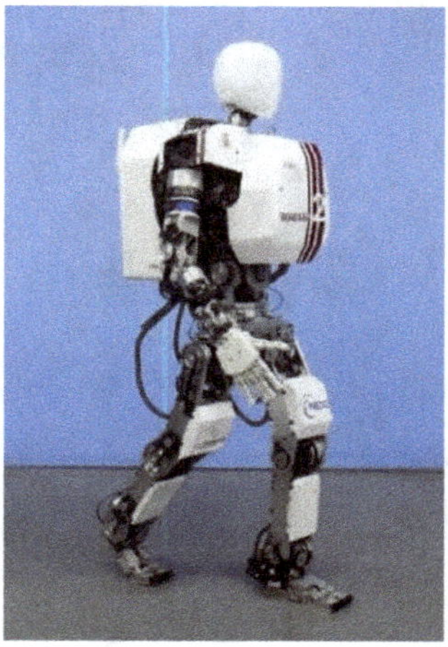

Figure 2-15 WABIAN-2R. [Courtesy of Ogura et al. [41].]

walking with knee stretching through adding DOFs of hip and passive toe joints [34]. The robot realized a quick turn using a slipping motion with both feet in 2011.

HRP-2 and HRP-3 were developed by AIST in 1998. These robots can walk stably, complete a variety of dexterous movements, carry objects with people, go across obstacles and lift objects from the ground. In order to increase the performance, HRP-4 (Figure 2-16) was developed with a light weight of only

Figure 2-16 HRP-4. [Courtesy of Kaneko et al. [42].]

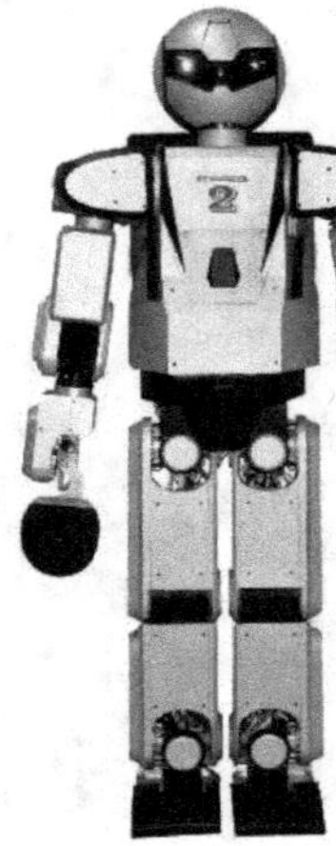

Figure 2-17 BHR-5 robot.

38 kg [35]. In 2012, HRP3L-JSK was developed by AIST and the University of Tokyo. It remains stable under external force while standing and can walk at a speed of 5 km/h [36, 37].

BHR series robots have been developed since 2000 by the Beijing Institute of Technology [38–40]. BHR-5 (Figure 2-17) is a newly built biped robot with vision, hearing, force and equilibrium sensing capability. It can walk straight and sideways, perform "Taiji" and martial arts, go around, climb stairs, play table tennis without external cables, and overcome disturbance. Some other famous joint position control biped robots, such as the LOLA [41], developed by the Technical University of Munich, Partner [42], developed by Toyota, and the KHR series and HUBO [43, 44], developed by KAIST, have also achieved significant progress during recent decades.

- Force control biped robots

  With the increasing requirements of safety, interaction and disturbance rejection ability of biped robots, many researchers focus on force/torque control robots. There are three driven modes for force control robots, namely electric, hydraulic and pneumatic. Hydraulic and pneumatic components drive joints directly while motors usually drive joints through series springs to achieve force control. This kind of robot can guarantee the compliance as well as the tracking precision.

Figure 2-18 Atalas. [Courtesy of Boston Dynamics.]

The biped robot PETMAN, developed by Boston Dynamics, is really amazing, both for its life-like appearance and powerful strength. It is designed for testing chemical protection clothing. It can realize natural agile movement, do push-ups, and go across obstacles. Boston Dynamics' newly built robot Atlas (Figure 2-18) is a high mobility biped robot designed to negotiate outdoor, rough terrain. In extremely challenging terrain, Atlas can

Figure 2-19 Sarcos. [Courtesy of Yamane et al. [81].]

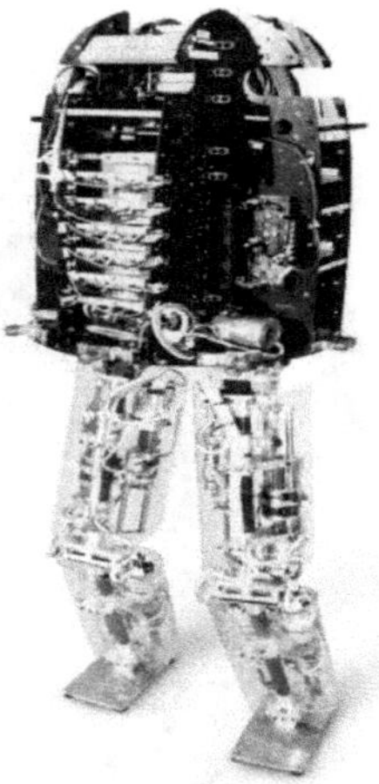

Figure 2-20 MVM2. [Courtesy of Pratt J et al. [56].]

go through congested space by using hands and feet [45]. Sarcos (Figure 2-19) is a commercial hydraulic drive robot with 51 DOFs. Many institutes including the Carnegie Mellon University, Japan JST and more experiments on this platform [46, 47].

M2V2 (Figure 2-20) was developed by IHMC with the aim of stability control under disturbance. It has 12 actuated DOFs in the lower limbs, which are powered by force controllable SEAs. These actuators provide high fidelity and low impedance, allowing for control techniques that exploit the natural dynamics of robots [48, 49]. Some other robots such as Spring Flamingo [50] developed by MIT are also force control robots.

In order to make robot motion more human-like, many force control robots are developed from the aspect of bionics. Robot Lucy, developed by the University of Brussels, was built with pneumatic artificial muscles, which could simultaneously control the torque and stiffness of joints in 2007. This type of control is similar to human muscle [51–53]. Tokyo University recently developed a robot, Kenshiro, as a real human simulator. It has the same joint arrangements and muscle as human, but the complex musculoskeletal construction make it difficult to control [54, 55].

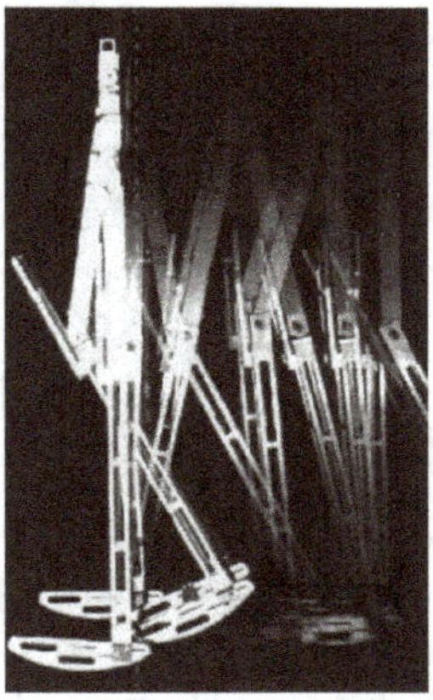

Figure 2-21 McGeer robot. [Courtesy of McGeer et al. [35].]

- Passive walking biped robots
  Passive dynamic walkers, pioneered by the works of McGeer in 1990 (Figure 2-21), present a fundamentally different type of walking machine. In these machines, simple mechanical principles are exploited for designing a mechanism which allows for a stable limit cycle. Then the concepts of those originally purely passive mechanism systems are used in dynamic limit-cycle walking machines, in which actuators are integrated to influence the limit cycle. Passive walking biped robots often show advantages in terms of human-like gaits and power consumption [56–58].

Figure 2-22 Flame. [Courtesy of Hobbelen et al. [60].]

Delft University of Technology developed their first generation of passive walking robots, Stappo, in 1995. Then they developed robot Denise in 2004, with a height of 1.5 m and a weight of 8 kg. The robot's hip is actuated by pneumatic power and the tilting and steering design of its ankle allows the robot to keep good balance at a walking speed of 1.4 km/h [59, 60]. In 2008 they developed a robot, Flame (Figure 2-22), with a height of 1.3 m, a weight of 15 kg, and a walking speed of 1.6 km/h. This robot has some sensors and seven motors. It can recover from tumbling through adjusting the foothold [61]. A battery powered robot with adult height, Ranger, was developed by Cornell University [62]. MIT's semi-passive robot Toddler (Figure 2-23) has no knees. It can start, stop, turn, go forward and backward through learning algorithms, and adapt to changing environments [63, 64].

2.2.2 Mechanism design

Mechanical structure is a key feature of biped robots. Many ingenious mechanism designs have been developed to enhance the performance of bipedal walking and mimic the structure of human. This part introduces mechanism design of feet, waist and ankle joints. Further readings about mechanism design principles can be found in Chapter 3 of this book.

- Foot design

  To realize stable, robust, natural and efficient walking, many foot mechanism designs were proposed. ASIMO [65] and HRP-2 [66] have impact absorption mechanisms in the feet. Since

Figure 2-23 Toddler. [Courtesy of MIT.]

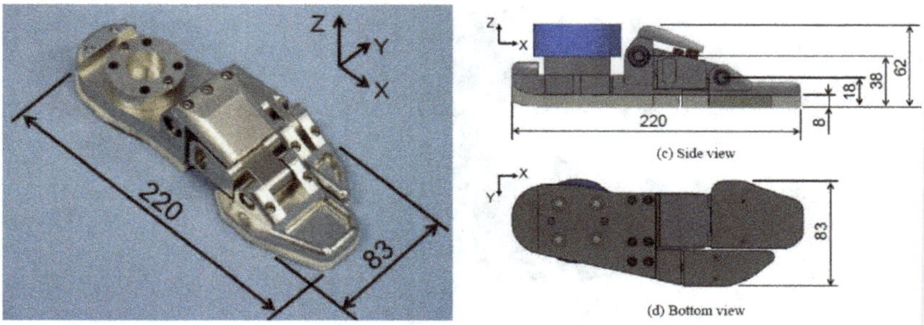

Figure 2-24 Foot mechanism structure. [Courtesy of Hashimoto et al. [69].]

biped robots with flat feet cannot perform heel strike and toe off, LOLA [67] and HRP-4C [68] have active toe joints to help realize a human-like walking, and extend the step lengths and active capabilities of the whole body.

Recently, a shoe-wearable foot mechanism which mimics a human's foot arch structure was designed by Waseda University to reveal the role of the human foot and absorb a foot-landing force at the plantar contact phase [69]. This structure is shown in Figure 2-24.

This shoe-wearable foot with a passive toe joint mechanism is 0.85 kg, and its size and the position of each joint axis are determined according to an adult's feet. Human foot arch structure and plantar characteristics are approximated through a rotational spring and compression springs, respectively, and the characteristics of human foot skin are mimicked by using a super-soft urethane resin called Hitohada gel [69].

- Waist design

A human's waist is a typical hyper-redundant system. It has a high level of flexibility, and it is the central regulation unit of a human body.

The waist of a biped can not only help to complete human-like motions, such as pitching, rolling and turning, but also be capable of regulating the center of gravity and compensating the yaw torque in order to keep balance. To expand the work space and realize the yaw torque compensation, 2-DOF waist with yaw and rotation joints (Figure 2-25) was developed by Waseda University [70].

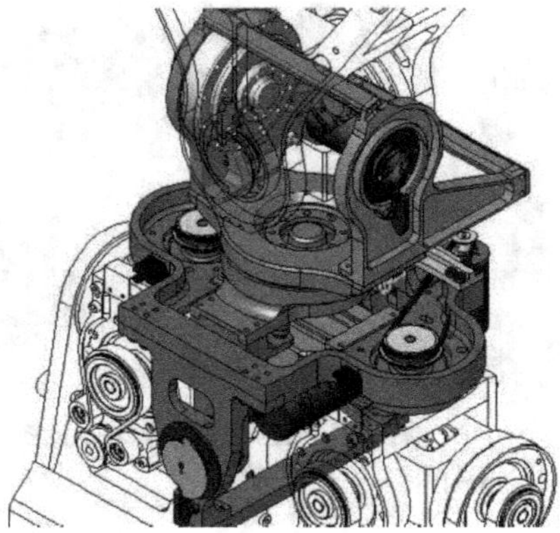

Figure 2-25 2-DOF waist structure. [Courtesy of Omer et al. [70].]

This waist joint allows the robot to roll and yaw independently, which allows it to increase step length and provides more mobility for lateral motions. The roll axis facilitates walking with a straight stance leg at a nearly constant height of the center of gravity [71]. The yaw axis movements can be used to compensate for the yaw torque between foot and ground while walking fast [72]. Some other robots like LOLA and BHR also have a similar waist joint configuration [73].

- Ankle design

In order to accomplish a better mass distribution in legs, large workspace and performance improvement in stabilizing control, robot LOLA was designed with a 2-DOF parallel mechanism in the ankle joint [73]. The kinematic actuation principle of the ankle joint is shown in Figure 2-26. The drives are designed as length variable steering rods in order to reduce manufacturing effort and part diversity. In addition, since ball screws are free from starting torque and have very low rolling friction, they guarantee precise control of foot torques that act on the ground.

As is shown in Figure 2-27, the ankle joint (3) is actuated by two spatial slider-crank mechanisms (7–9). The motors (4) are mounted on the thigh (1), close to the hip joint. Each linear drive is composed of a roller screw (8) and a linear bearing (9). The roller screw is arranged in longitudinal direction of the shank

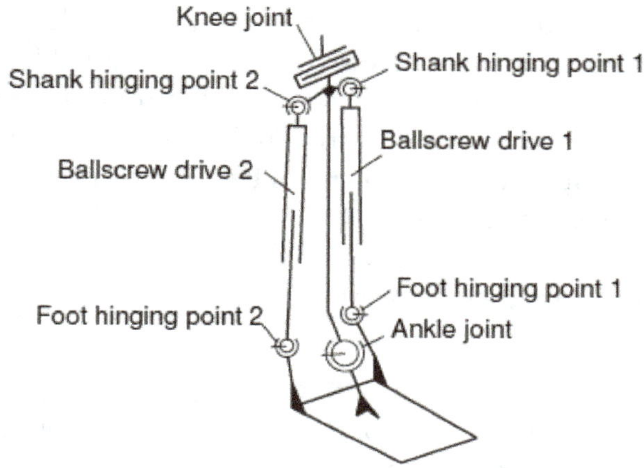

Figure 2-26 Kinematic actuation principle of ankle joint. [Courtesy of Ulbrich et al. [73].]

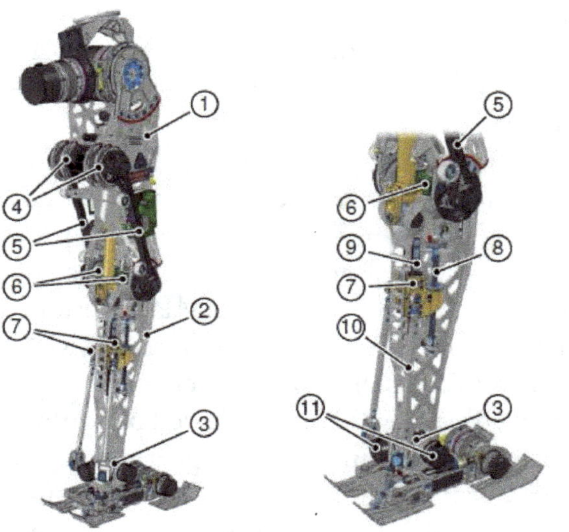

1	Thigh	5	Synchronous belt	9	Linear guide
2	Shank	6	Bevel gear	10	Steering rod
3	Ankle joint	7	Linear carriage	11	Angular sensor
4	Motor	8	Roller screw		

Figure 2-27 2-DOF parallel mechanism in the ankle joint. [Courtesy of Lohmeier et al. [67].]

and perpendicular to the knee joint axis, and the linear bearing is utilized to keep the screw free from the radial loads. The foot is linked with the linear carriage (7) through the steering rod (10). The bevel gear (6) in the knee joint axis is actuated by the motor through the synchronous belt (5). And the output shaft of the bevel gear drives the roller screw. Finally, the movement of the ankle joint is achieved. The absolute angular sensors (11) can measure joint angles directly.

2.2.3 Motion planning

It is challenging work to plan biped walking motion due to the nonlinear dynamics and high coupling of the robot system. ZMP criterion and 3D linear pendulum are the basic methods to design the motion for a biped robot, and more natural walking can be obtained through analysis of human motion data.

- ZMP criterion
  Zero Moment Point (ZMP) criterion is a widely adopted principle in motion planning for bipedal walking, which was presented in 1969 by Vukobratović [74]. The ZMP is defined as the point on the ground around which the sum of all the moments of active forces equals zero. If ZMP is within the convex hull of all contact points between the feet and the ground, the biped robot is able to walk. In order to ensure dynamic stability for a biped robot, one common method in motion planning for the traditional biped robot based on ZMP is to design a desired ZMP trajectory that strictly satisfies "ZMP criterion," then derive the whole body motion to realize the desired ZMP trajectory.
- 3D linear inverted pendulum model
  Kajita et al., used the 3D liner inverted pendulum model (LIPM) to generate a pattern in single support phase (SSP) and to keep the velocity of the center of mass (CoM) constant in double support phase (DSP) [75]. They were the pioneers who realized stable walking using LIPM. In conventional LIPM, as shown in Figure 2-28, it is assumed that the mass of the robot can be lumped to the CoM, which is supported by a massless telescopic leg. Y axis points to the walking direction. X axis points to the right of the model. Z axis is vertical. When the CoM is kept on

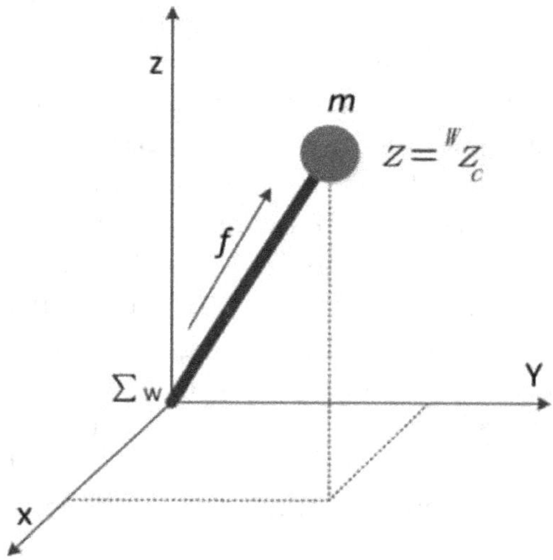

Figure 2-28 3D liner inverted pendulum model.

a constant height, its motion in the frontal direction is independent of the motion in the lateral direction. With the assumption that $z = z_c$, the dynamic equations are given as follows:

$$\ddot{x}_c - \frac{g}{z_c} x_c = -\frac{g}{z_c} x_{zmp} \tag{7}$$

$$\ddot{y}_c - \frac{g}{z_c} y_c = -\frac{g}{z_c} y_{zmp} \tag{8}$$

where g is gravity acceleration, (x_c, y_c) is the CoM coordinate in the world coordinates, and (x_{zmp}, y_{zmp}) is the ZMP coordinate in the world coordinates.

For further simplification, ZMP is fixed in many methods. And so x_{zmp} and y_{zmp} equal zero if the origin point of the coordinate is set to the support point. In this case, Equations (7) and (8) become easier and an analytic expression of the CoM motion can be obtained [75].

- Multi-link model

The multi-link model has higher accuracy of dynamic and kinematic compared with the simple models. Pre-design ZMP trajectories and pre-design walking motion are two main methods of this model. The first one is introduced in this section, and the other will be described in Section of 2.2.6.

The research on the stable walking pattern generation was initiated in Waseda University [76]. Takanishi et al. designed the reference ZMP and realized continuous biped walking with the trajectories of lower and upper body. The walking pattern composed of unit patterns of the foot and the waist is first designed by a polynomial according to the ground conditions. Then the compensatory unit patterns of the waist and trunk are calculated to achieve more stable walking. Kinematic constraints are imposed on the upper body motion to solve the interferential and nonlinear differential equation. The left and right side of the

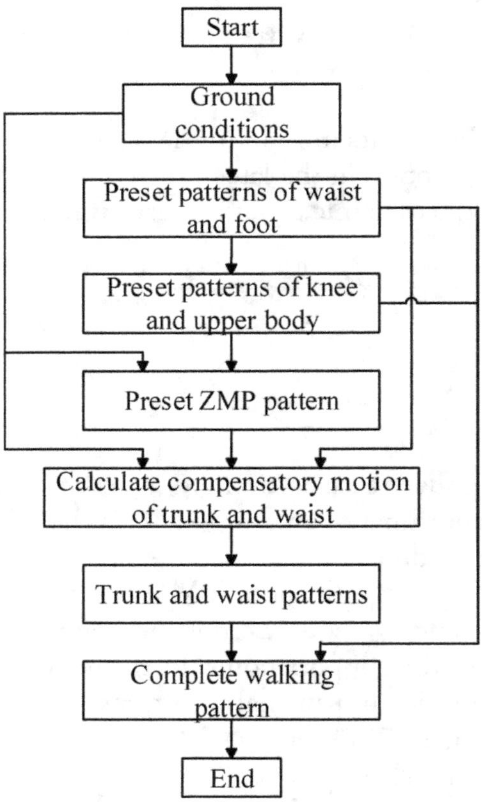

Figure 2-29 Walking pattern generation based on ZMP. [Courtesy of Lim et al. [76].]

equation is regarded as the periodic function, which is solved by comparing the coefficients of both sides of the equation using fast Fourier transform (FFT). And a more accurate solution could be achieved through iteration. The knee unit pattern is geometrically computed on the basis of the foot pattern and the compensatory waist pattern. Finally, the complete continuous walking pattern that consists of these unit patterns is obtained according to the step direction, as shown in Figure 2-29.

2.2.4 Stability control

Stability control is also an essential part in the development of biped robots, and research has proposed various solutions to this issue.

Choi et al., introduced a controller with formula stability proof using the Lyapunov method [77]. Wieber focused on compensating strong perturbations of robots and proposed methods to update the upper body trajectory using a linear model predictive control scheme [78]. Takenaka et al., focused on handling disturbance through upper body position feedback gains design and the extended model ZMP control method [79]. Sugihara evaluated the stable pole of an inverted pendulum which could extend the admissible region of ZMP [80]. Kajita et al., proposed a CoM/ZMP tracking controller based on the linear inverted pendulum, considering a ZMP delay [81]. Pratt et al., presented the capture point which can lead the COG motion to a complete stop [82]. Englsberger et al., proposed and proved the derivation of a capture point based on the natural dynamics of the linear inverted pendulum (LIP), which stabilizes walking robots [83].

The biped robot may suddenly become unstable and begin to tip over when there are unexpected sudden events. In that case, ZMP becomes uncontrollable, since the contact force between the ground and the feet cannot provide the necessary recovery moment to control the ZMP. For example, in the case when the actual ground surface is lower than the assumed surface for the planned dynamic pattern, the swing foot will land on the ground more slowly than the designed time of the planned dynamic pattern, which results in the biped tipping forward immediately. In addition, when the biped becomes unstable, the biped body slants from the desired body inclination. Thereafter, the body inclination is called the body posture. If the body posture is not recovered in time, the tipping moment becomes large and the body slanting increases continuously, which results in the robot tipping fast. Control methods based on sensory reflex are the general ways to solve this problem, as they not only control the ZMP, but also the parameters of the landing time and the body posture online are proposed [84].

The stability control based on sensory reflex presented in this section consists of a ZMP reflex, a landing-phase reflex and a body-posture reflex. The three sensory reflexes are independent, and become active when their conditions are satisfied. The assumptions for the sensory reflex are given as follows:

1) The unexpected irregularities can be detected by the biped sensors;
2) The noise of sensors used for the sensory reflex can be filtered, and their responses are fast enough for the real-time control.

- ZMP reflex

  The ZMP reflex controls the ZMP of actual walking online. The actual ZMP is measured by a foot-force sensor. Generally, only one edge of the foot sole is in contact with the ground during biped tipping. For example, only the toe is in contact with the ground when the biped tips forward. Viewed from the perspective of the ZMP, in those cases, the actual ZMP is on the toe; that is, the actual ZMP is on the boundary of the stable region. To avoid having only one edge of the foot sole contacting the ground, there should be some distance between the actual ZMP and the boundary of the stable region. That is the stability margin.

  If the actual ZMP is always controlled in the center of the stable region, the actual stability margin is the largest. But in that case, the robot cannot move at high speed. To keep walking in the most dynamic pattern possible, the actual stability margin is specified to be the same as the stability margin of the dynamic pattern. Thereafter, the region with such a stability margin is defined as the desired stable region (Figure 2-30).

  One effective way to keep the actual ZMP within the desired stable region is to control the ankle joints of the support feet (Figure 2-31). In the case when the actual ZMP is within the desired stable region, the ZMP reflex is not active. In the case when the actual ZMP is outside the desired stable region, the ZMP reflex is active, and the ankle joint $\Delta\theta_a(nT_s)$ realizing the ZMP reflex is calculated using the following equation:

$$\Delta\theta_a(nT_s)=\sum_{j=1}^{j=n}\delta\theta_a(jT_s)$$
$$\delta\theta_a(jT_s)=\begin{cases} K_{ac}d_{zmp}(jTs), & F_z>0 \\ -K_{as}\Delta\theta_a((j-1)T_s), & F_z=0 \end{cases} \tag{17}$$

Where T_s is the sample period of the servo loop, and nT_s is the current time, $d_{zmp}(jTs)$ is the distance between the actual ZMP and the boundary of the desired stable region, K_{ac} and K_{as} are coefficients, and F_z is the foot contact force with the ground, which is measured by the sensor. If $F_z > 0$, that is, the foot is in contact with the ground, the ankle joint of this foot is used to control the actual ZMP.

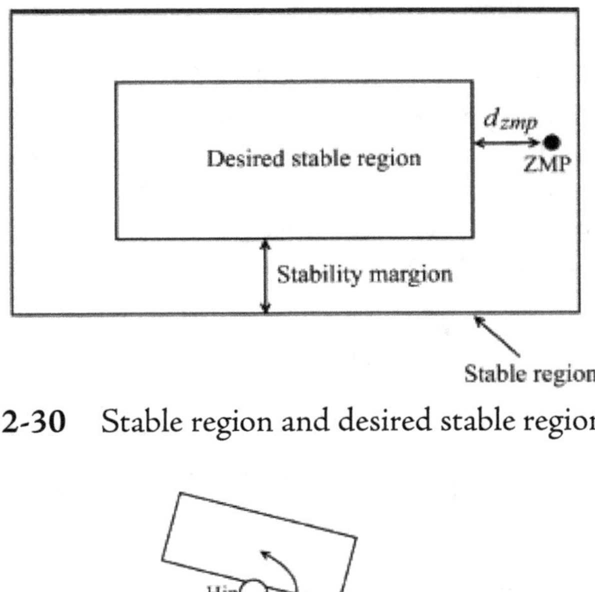

Figure 2-30 Stable region and desired stable region.

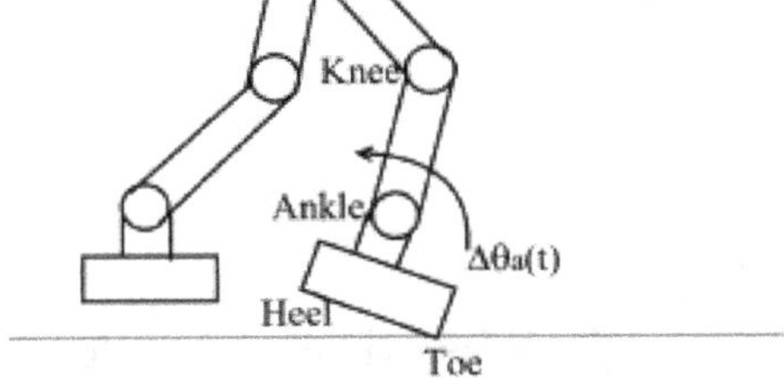

Figure 2-31 ZMP reflex and body-posture reflex.

If $F_z = 0$, the foot becomes the swing foot, and the ankle joint of this foot returns to the value of its dynamic pattern gradually.

- Landing-phase reflex

The biped may suddenly become unstable if the swing foot cannot land on the ground in time due to unexpected ground irregularities. For example, the ground surface is assumed to be level when the dynamic walking pattern is generated, but the actual ground surface is unknown rough terrain. In the case when the actual ground surface is higher than the assumed surface of the planned dynamic walking pattern (Figure 2-32(a)), the swing foot lands on the ground faster than the designed time of the planned dynamic walking pattern. Then the moment of tipping backward occurs, and the robot will tip backward. Conversely, the foot lands on the ground too slowly when the actual ground surface is lower than the assumed surface (Figure 2-32(b)), the moment of tipping forward increases, and the robot will tip forward.

The landing-phase reflex controls the swing foot landing on the ground at a desired time online. In the case when the swing foot lands on the ground too fast, the robot lifts its foot. The increase of foot height $\Delta Z_f(nT_s)$ to realize the landing phase reflex is given as follows:

$$\Delta Z_f(nT_s) = \sum_{j=1}^{j=n} \delta Z_f(jTs)$$
$$\delta Z_f(jT_s) = \begin{cases} K_{fc}F_z, & 0 < F_z < M_r \\ -K_{fs}\Delta Z_f((j-1)T_s), & F_z \geq M_r \end{cases} \tag{18}$$

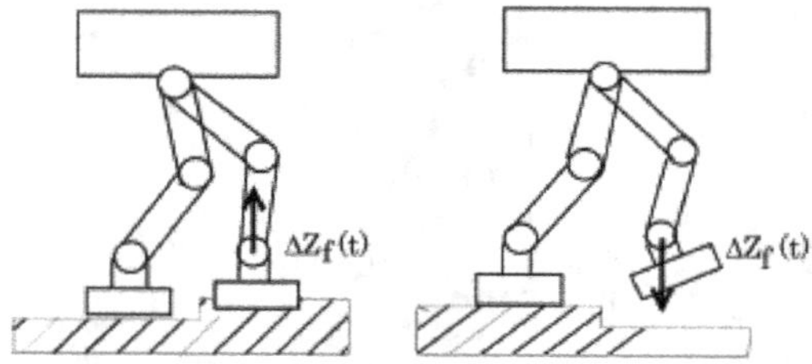

(a) Landing too fast (b) Landing too late

Figure 2-32 Landing-phase reflex of foot position.

where K_{fc} and K_{fs} are coefficients, and M_r is an approximate value of the weight of the biped. If $F_z \geq M_r$, the support foot nearly becomes the single support foot, then the foot height returns to the value of the dynamic walking pattern. In the case when the swing foot lands on the ground too late, the robot lowers its foot to land on the ground. The increase of foot height $\Delta Z_f(nT_s)$ to realize the landing-phase reflex is given as follows:

$$\begin{aligned} \Delta Z_f(nT_s) &= \sum_{j=1}^{j=n} \delta Z_f(jTs) \\ \delta Z_f(jT_s) &= \begin{cases} -h, & F_z = 0 \\ -K_{fs}\Delta Z_f((j-1)T_s), & F_z \geq M_r \end{cases} \end{aligned} \tag{19}$$

where h is a constant variable. The online change of foot height $Z_f(nT_s)$ is achieved by controlling the hip and knee joints according to the kinematics.

- Body-posture reflex
  From the viewpoint of stability, it is desirable that the body posture is constant when there is no waist joint. The actual body posture is measured by inclination sensors, such as accelerometers and angular rate sensors. The actual body posture will slant from its desired body posture when the biped begins to tip over. As the slanting increases, the tipping moment becomes larger. The body-posture reflex controls the actual body posture to be the desired body posture online. Since the hip joints are near the body, the most effective way to keep the desired body posture is to control the hip joints (Figure 2-32). The hip joint $\Delta\theta_h(t)$ to realize the body-posture reflex is given as follows:

$$\begin{aligned} \Delta\theta_h(t) &= \sum_{j=1}^{j=n} \delta\theta_h(jT_s) \\ \delta\theta_h(jT_s) &= \begin{cases} \Delta\theta_{actb}(jT_s), & F_z > 0 \\ -K_{hs}\Delta\theta_h((j-1)T_s), & F_z = 0 \end{cases} \end{aligned} \tag{20}$$

where $\Delta\theta_{actb}(jT_s)$ is the deflection between the actual body posture and the desired body posture, and K_{hs} is the coefficient. If

$F_z > 0$, the foot is in contact with the ground, and the hip joint of this foot is used to control the body posture. On the other hand, if $F_z = 0$, the foot is the swing foot, and its hip joint does not affect the body posture. In this case, the hip joint of the swing foot returns to the value of the dynamic walking pattern.

2.2.5 Control system

The main architecture for biped robots can be summarized as follows. First, some biped robots, such as P3 [85], HRP-2 [86] and BHR-01 [87], adopt the centralized control architecture based on local VME, ISA or PCI bus. The advantage of this architecture is high communication speed and capacity. But the calculation burden of the central computer is so heavy that it could not process much other sensor information. Additionally, the wires connected to the central computer are excessive. The long wires easily suffer from electro-magnetic interference. In order to solve these problems, some other biped robots, such as HRP-3P [79], KHR-2 [88], QRIO [89], JOHNNIE [90] and BHR-03, adopt the distributed control architecture based on CAN, OPEN-R, USB or Ethernet. In this section, an open distributed control system for BHR-5 [91] is introduced, which improves the compatibility and expandability of the whole system, and also reduces the calculation burden of the central computer.

The open distributed control system for BHR-5, as shown in Figure 2-33, has three sub-systems: the teleportation system, decision-making system (DMS) and motion control system (MCS). The teleportation system is used to control the robot remotely. DMS is used to process non real-time events, like vision, hearing and decision-making. The operating system (OS) of these two sub-systems is Windows XP. MCS controls all joints in real-time. Its OS is RT-Linux. Wireless LAN and Ethernet are used as communication interface between them. Both DMS and MCS support the CANopen protocol, which facilitates integrating new peripherals.

The MCS includes a coordinated control computer, PCI cards, CANopen drivers, Mini Multi-Axis Controllers (MMAC), DC motors and various sensors.

Two 6-axis force/torque sensors are located on each foot to detect the contact force between the foot and the ground. In addition, two 6-axis

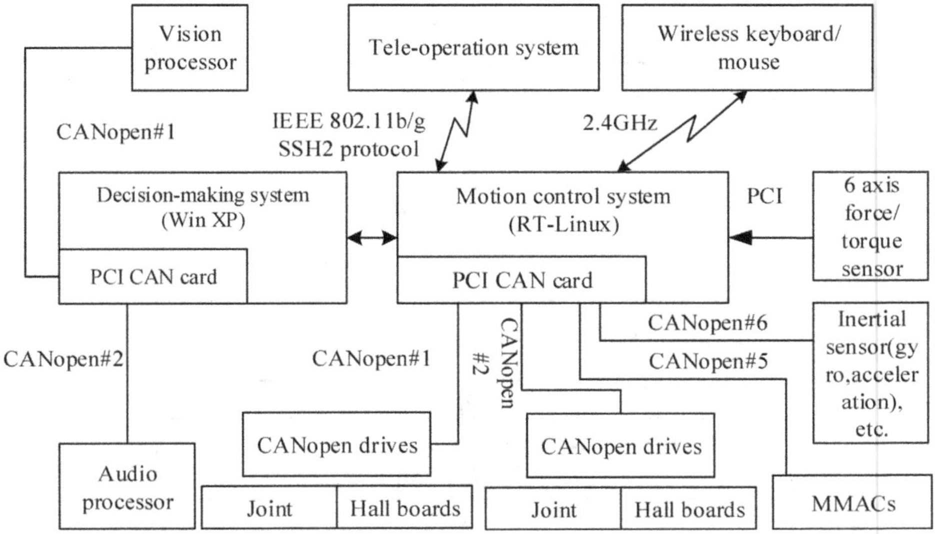

Figure 2-33 The open distributed control system.

force/torque sensors are located on each wrist. The force/torque information is processed by a 4-channel PCI interface card. In order to measure the attitude/posture of the upper body, inertial sensors are assembled in the body. Keyboard and mouse are use to control this computer via USB.

Three dual channel PCI CAN cards are assembled on the mainboard. Most of the peripherals communicate with the control computer via CAN bus. The highest baud rate of CAN bus is 1 Mbits/s, and sending a frame always costs 120 us. In order to enhance the operating speed, these channels are processed in the parallel mode. That is to say, when the PC sends CAN frames through one channel, it just writes a frame into its transmission buffer and does not wait for an acknowledgment signal that it was sent successfully. In succession, the PC writes a frame into another channel's buffer in the same way. When the PC receives frames, it reads receiving buffers of the six channels circularly. This method helps to save a great deal of time and the communication period can drop to less than 3 ms.

Figure 2-34 shows the flow chart of periodic servo control thread at a 5 ms servo period in MCS. The real-time control process routine operates sensory reflex control to generate trajectory reference data for all joints.

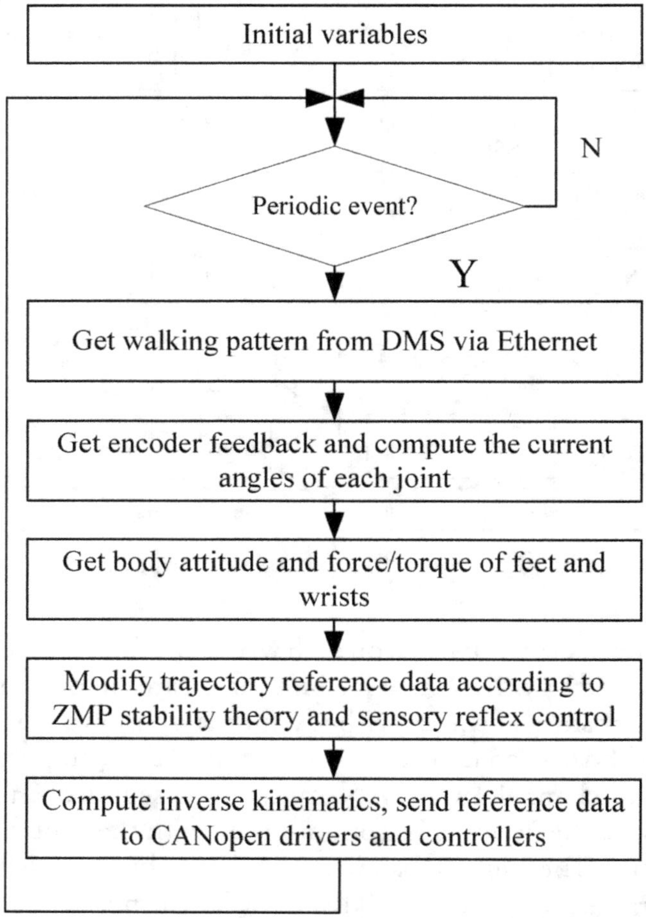

Figure 2-34 Flow chart of real-time periodic control.

2.2.6 Biped walking

In this section, a method of walking planning for BHR-5 considering human walking data and ZMP stability margin is introduced. Then, a walking experiment using stability control is presented.

- Description of BHR-5
  BHR-5 is a newly developed biped robot platform with torso, head, arms, hands, legs and waist. The height of this robot is 1.65 m and the weight is 60 Kg. In order to realize the stability control and cope with the perturbation of external environment,

six-axis force/torque sensors, accelerometers and gyroscopes are installed inside the robot. The main difference between BHR-5 and the previous editions is that two DOFs of waist, yaw and roll joints are added to the torso (Figure 2-35), which makes the movement of the robot more human-like.

- Human motion data analysis

Recently, some researchers have focused on the walking mechanism and characteristics of the human to make bipedal walking more natural and human-like, and the motion capture system was adopted to collect human walking data [92–100].

In order to analyze the human walking mechanism, a motion capture system is utilized to record the main joint angles of the human body. The system consists of 12 optical cameras and 38 reverberation markers that are pasted on the surface of a subject's body. It records the positions of all markers with the rate of 100 frames per second. And the main angles of joints can be

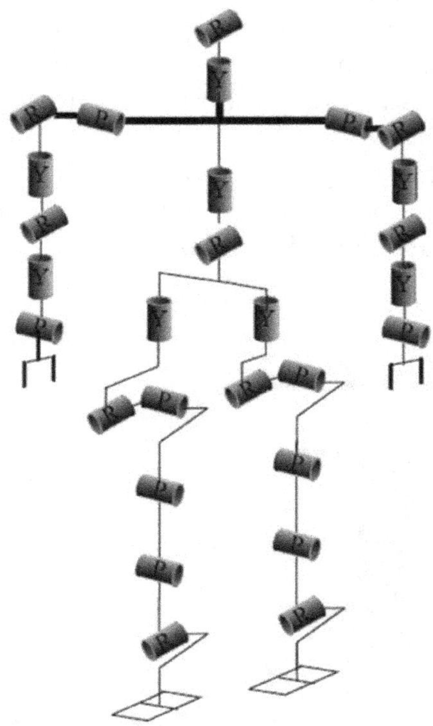

Figure 2-35 DOFs configuration of BHR-5.

calculated by software. Then human walking data with several step lengths and cycles can be captured from different subjects with different physical characteristics. Figure 2-36 shows the subject with reverberation markers and the skeleton structure formed in the software.

As a result of analyzing the captured walking data, the human walking characteristics discovered are [101]:

(1) A one-step cycle can be described by the DSP and the SSP. The rate of DSP (the ratio between the double support time and the whole step cycle) during a step cycle will be less if the step cycle is longer for the same step length, and vice versa. And the rate of DSP during a step cycle will be larger if the step length is shorter for the same step cycle, and vice versa.

(2) The swing height of the foot is relevant to step length, not to step cycle. The longer the step length is, the higher the height is. And the swing foot moves up to the highest position at approximately 3/5 the time of SSP. The slope angles of the swing foot as it leaves and lands on the ground will be larger if step cycle is shorter for the same step length, and vice versa.

(3) The rate of hip displacement in sagittal direction during DSP will be less if the step cycle is shorter for the same step length, and vice versa.

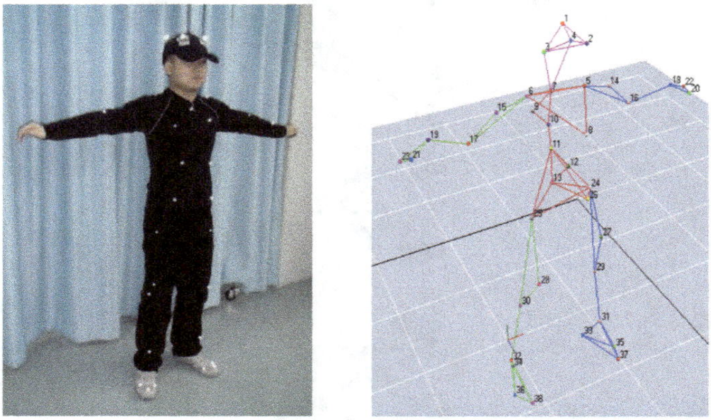

Figure 2-36 The subject and the skeleton structure.

(4) The hip displacement in lateral direction during DSP will be less if the step cycle is shorter for the same step length, and vice versa.
(5) The hip displacement in vertical direction is relevant to step length, not to step cycle. The longer the step length is, the larger displacement is. The hip moves up to the highest position at approximately 3/5 the time of SSP, and moves down to the lowest position at the middle time of DSP.
(6) The hip moves close to folding line in level plane. There is an approximate linear relationship between sagittal movement and lateral movement during double-support phase.

- Pattern generation
  The key parameters of foot trajectories and hip trajectories are planned considering the ground conditions and the characteristics of human walking parameters collected from humans. Then the final motion pattern is determined considering the stability margin. After that, joint angles can be obtained by inverse kinematics [92, 102]. Here, the stability margin is the minimum distance between the ZMP and the boundary of the supporting polygon.

  For the foot trajectories, supposing that the period necessary for one walking step is T, the K_{th} walking step begins with the heel of the foot touching the ground at $t = KT$, and ends with the heel of the foot touching the ground at $t = (K + 1)T$. The constraints of the foot based on human walking characteristics are given by the following equations (the walking parameters are shown in Figure 2-37):

$$\theta_f(t) = \begin{cases} q_s(K), & t = KT_c \\ q_b, & t = KT_c + T_d \\ q_f, & t = KT_c + T_c \\ q_b(K), & t = (K+1)T_c + T_d \end{cases} \tag{9}$$

Where $\theta_f(t)$ represent the angle of foot, T_c is the step period, T_d is the period of DSP, and T_d/T_c denotes the rate of DSP during

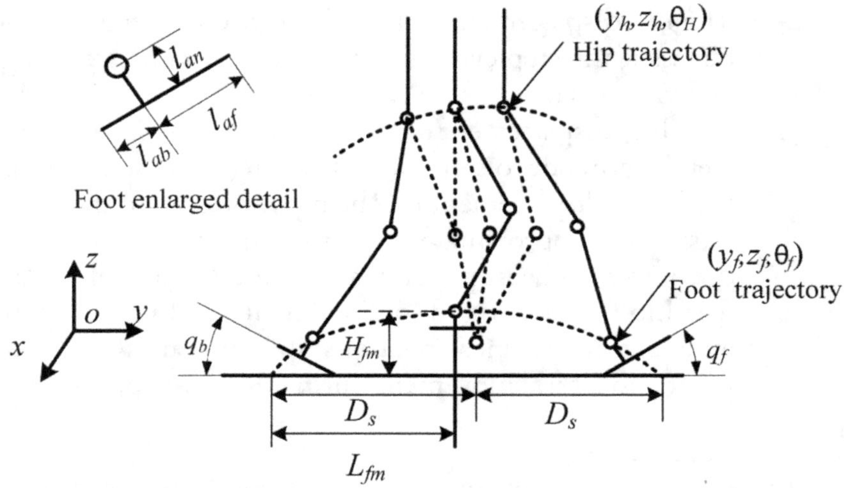

Figure 2-37 The walking parameters.

a step cycle, which can be assigned according to the human data. $q_s(K)$ and $q_e(K)$ are the angles of the ground surface under the support foot, particularly $q_s(K) = q_e(K) = 0$ on level ground, q_b and q_f are the slope angles of the foot as it leaves and lands on the ground. All of these parameters can be assigned based on the human walking mechanism and ground conditions:

$$y_f(t) = \begin{cases} KD_s, & t = KT_c \\ KD_s + l_{af}(1-\cos q_b) + l_{an}\sin q_b, & t = KT_c + T_d \\ KD_s + l_{fm}, & t = KT_c + T_m \\ (K+2)D_s - l_{ab}(1-\cos q_f) - l_{an}\sin q_f, & t = KT_c + T_c \\ (K+2)D_s, & t = (K+1)T_c + T_d \end{cases} \tag{10}$$

$$z_f(t) = \begin{cases} h_{gs}(K) + l_{an}, & t = KT_c \\ h_{gs}(K) + l_{an}\cos q_b + l_{af}\sin q_b, & t = KT_c + T_d \\ H_{fm}, & t = KT_c + T_m \\ h_{ge}(K) + l_{an}\cos q_f + l_{ab}\sin q_f, & t = KT_c + T_c \\ h_{ge}(K) + l_{an}, & t = (K+1)T_c + T_d \end{cases} \tag{11}$$

where $y_f(t)$ $z_f(t)$ represents the foot coordinates in y and z directions, respectively. D_s is the step length, l_{an} is the height of ankle joint, l_{af} denotes the length from the ankle joint to the toe, and l_{ab} denotes the length from the ankle to the heel. (l_{fm}, h_{fm}) is the coordinate of the highest point of the swing foot, which can also be assigned according to human data. $h_{gs}(K)$ and $h_{ge}(K)$ are the heights of the ground surface under the support foot, particularly $h_{gs}(K) = h_{ge}(K) = 0$ on level ground. Hence, the foot trajectory can be obtained based on the ground conditions, and the mechanism and data collected from humans.

In the vertical direction, the waist moves up and down, according to the obtained human walking parameters, and the following equation should be satisfied:

$$z_h(t) = \begin{cases} H_{h\min}, t = KT_c + 0.5T_d \\ H_{h\max}, t = KT_c + 0.6(T_c - T_d) \\ H_{h\min}, t = (K+1)T_c + 0.5T_d \end{cases} \tag{12}$$

where $H_{h\min}$ and $H_{h\max}$ are hip displacement at the lowest and highest positions in the vertical plane, respectively.

$y_h(t)$ can be described as follows:

$$y_h(t) = \begin{cases} kD_s + y_{sd}, & t = KT \\ kD_s + y_{md}, & t = KT + T_d/2 \\ (k+1)D_s + y_{ed}, & t = (K+1)T \end{cases} \tag{13}$$

where y_{sd}, y_{md} and y_{ed} are the hip displacement at the beginning, middle and end of the DSP in sagittal plane, respectively. And the dispersion between y_{ed} and y_{sd} can be defined according to the human walking data.

From the human walking characteristics above, there is an approximate linear relationship between sagittal movement and lateral movement during DSP, so the following constrains should be satisfied:

$$J=\sum_{i=1}^{N}[y_i-(a+bx_i)]^2<\varepsilon \quad \& \quad \begin{cases} \dfrac{\partial J}{\partial a}=0 \\ \dfrac{\partial J}{\partial b}=0 \end{cases} \tag{14}$$

where y_i are the points in sagittal hip trajectory, x_i are the points in the lateral hip trajectory and ε is the threshold.

By using third spline interpolation, the trajectory of $y_h(t)$ that satisfies the constraints equation (13) and the constraint of second derivative continuity can be obtained. Furthermore, $x_h(t)$ can be considered as the function of $y_h(t)$, then trajectory of $x_h(t)$ that satisfies the constraints equation (14), and the constraint of second derivative continuity can be obtained by using third order spline interpolation. Hence, the series of smooth $x_h(t)$ and $y_h(t)$ are obtained by setting different values of y_{sd} and y_{ed} with fixed ranges.

After the trajectory design of foot and hip, the final trajectory is determined by taking the largest stability margin into account. The ZMP can be computed by the following equations:

$$x_{zmp}=\frac{\sum_{i=1}^{n}m_i(\ddot{z}_i+g)x_i-\sum_{i=1}^{n}m_i\ddot{x}_iz_i-\sum_{i=1}^{n}I_{iy}\ddot{\Omega}_{iy}}{\sum_{i=1}^{n}m_i(\ddot{z}_i+g)} \tag{15}$$

$$y_{zmp}=\frac{\sum_{i=1}^{n}m_i(\ddot{z}_i+g)y_i-\sum_{i=1}^{n}m_i\ddot{y}_iz_i-\sum_{i=1}^{n}I_{ix}\ddot{\Omega}_{ix}}{\sum_{i=1}^{n}m_i(\ddot{z}_i+g)} \tag{16}$$

Where $(x_{zmp}, y_{zmp}, 0)$ is the coordinate of the ZMP, n is the number of links, m_i is the mass of link i, g is the gravitational acceleration, (x_i, y_i, z_i) is the coordinate of the mass center of link i on the absolute Cartesian coordinate system, I_i is the rotational inertia of link i, and $\ddot{\Omega}_i$ is the angle acceleration of link i.

Finally, the smooth hip trajectories with the largest stability can be formulated through selecting the y_{sd} and y_{ed}, and the process of the improved method of walking pattern generation is shown in Figure 2-38.

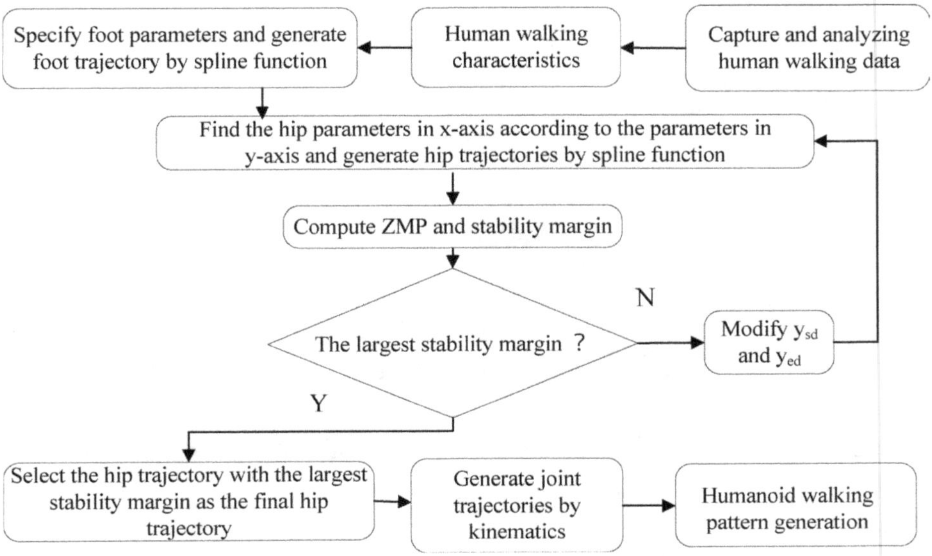

Figure 2-38 The flow chart of walking pattern generation.

- Simulation and experiment

  Biped walking is a complex problem involving mechanism design, motion planning and stability control. In this section, the walking simulation and experiment of BHR-5 using the theory mentioned above is illustrated. The parameters of the planning trajectory are in Table 2-1, and the rate of DSP is 20%, which is inspired from human data.

  Figure 2-39 is the stick figure of walking, including 6 steps. It shows that the feet leave and land on the ground with desired angles.

Table 2-1 Walking parameters.

Walking speed	1.5 km/h
Walking stride	0.33 m
Walking cycle	0.8 s
Single-supporting phase duration	0.16 s
Double-supporting phase duration	0.64 s
Steps number	6

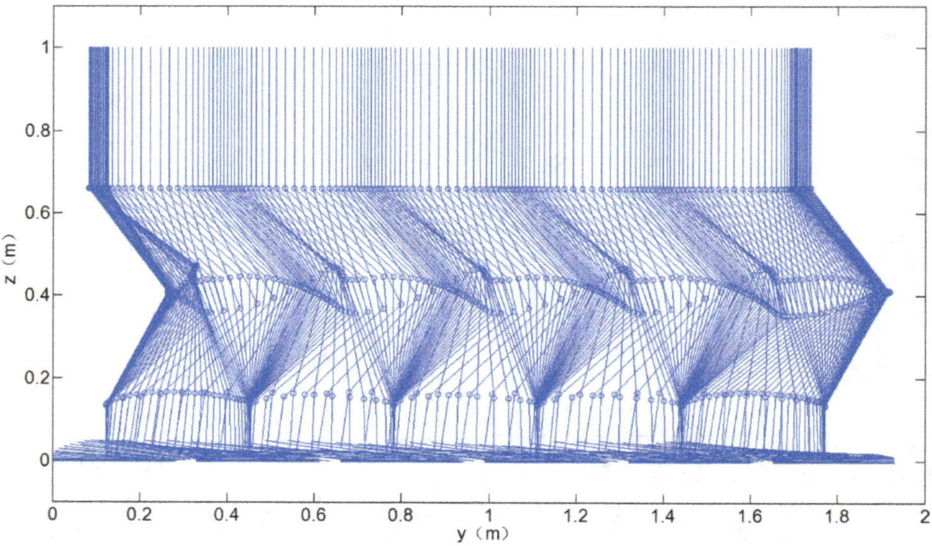

Figure 2-39 Stick figure of walking.

Figure 2-40 shows the ZMP in X and Y direction. The ZMP is always in the convex hull and the robot is able to walk according to the "ZMP criterion." Figure 2-41 shows the experiment of two steps. It shows that BHR-5 walks on the ground steadily with the speed of 1.5 km/h.

2.3 Conclusions

This chapter introduces the history and key technologies of exoskeleton robots and biped robots. On the one hand, research into exoskeleton robots and biped robots has made great achievements, especially in key issues concerning motion planning, flexible control, and balance control. On the other hand, several exoskeleton and biped robot platforms have been built and used in a certain environment.

There are still many issues needed to be resolved in this subject. The applications and safety control are key issues which require subsequent progresses in collaboration among numerous disciplines, including biology, robotics, mechanics, electronics, et al. With the development of related technologies, exoskeletons and bipeds will bring a revolution to every aspect of our life.

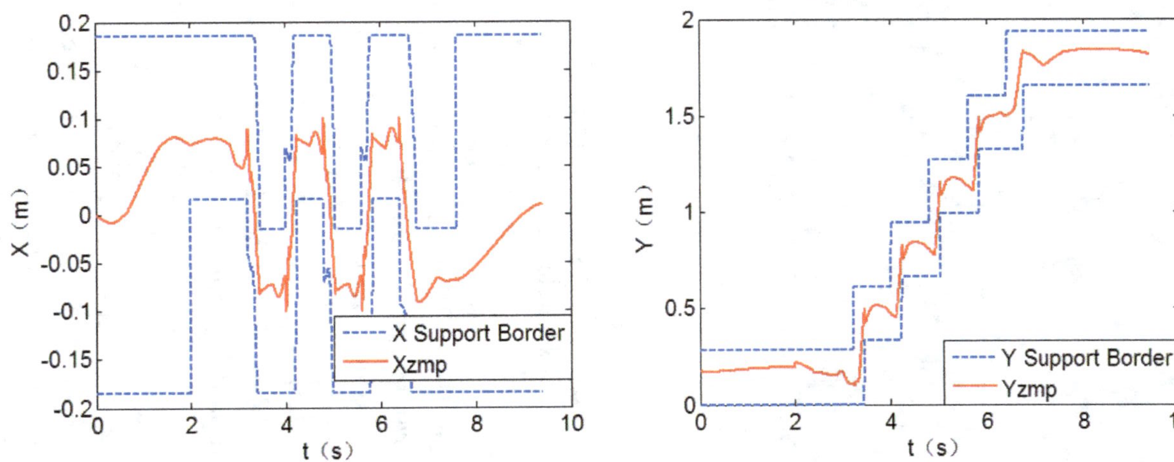

Figure 2-40 ZMP in X and Y direction.

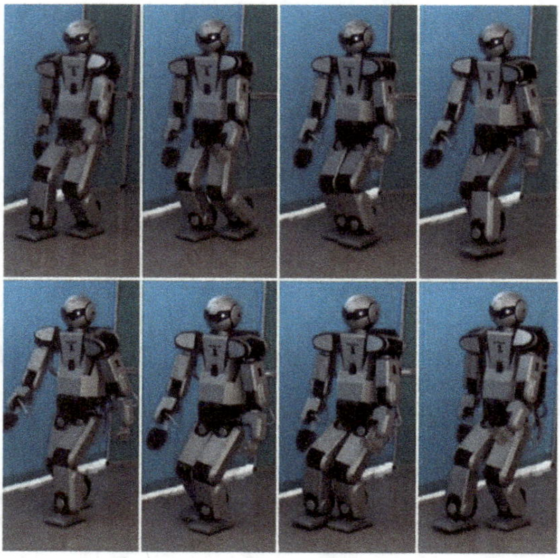

Figure 2-41 Walking experiment.

2.4 References

[1] Rosen, J., Brand, M., Fuchs, M. B., Arcan, M., 2001, A myosignal-based powered exoskeleton system. *IEEE Transactions on Systems, Man and Cybernetics*, **31**(3), pp. 210–222.

[2] Kazerooni, H., 2005, Exoskeletons for human power augmentation, *IEEE/RSJ International Conference on Intelligent Robots and Systems*, pp. 3459–3464.

[3] Perry, J. C., Rosen, J., Burns, S., 2007, Upper-limb powered exoskeleton design, *IEEE/ASME Transactions on Mechatronics*, **12**(4), pp. 408–417.

[4] Gilbert, E., Callan, C., 1968, Hardiman I prototype, General Electric Company, Schenectady, NY, GE Tech. Rep. S-68-1081.

[5] Makinson, B. J., 1971, Research and development prototype for machine augmentation of human strength and endurance, *Hardiman I Project* (No. S-71-1056).

[6] *Springer handbook of robotics*, 2008. Springer.

[7] Hardiman I prototype project, 1968, special interim study, General Electric Co., Schenectady, NY, Rep. S-68-1060.

[8] Zoss, A., Kazerooni, H., Chu, A., 2005, On the mechanical design of the Berkeley Lower Extremity Exoskeleton (BLEEX),

2005 IEEE/RSJ International Conference on Intelligent Robots and Systems, pp. 3465–3472.

[9] Kawamoto, H., Lee, S., Kanbe, S., Sankai, Y., 2003, Power assist method for HAL-3 using EMG-based feedback controller, *IEEE International Conference on Systems, Man and Cybernetics*, **2**, pp. 1648–1653.

[10] Kiguchi, K., Tanaka, T., Fukuda, T., 2004 Neuro-fuzzy control of a robotic exoskeleton with EMG signals, *IEEE Transactions on Fuzzy Systems*, **12**, pp. 481–490.

[11] Zoss, A., Kazerooni, H., Chu, A., 2005, On the mechanical design of the Berkeley Lower Extremity Exoskeleton (BLEEX), *2005 IEEE/RSJ International Conference on Intelligent Robots and Systems*, pp. 3465–3472.

[12] Nicolelis, M. A., 2012, Mind in motion. *Scientific American*, **307**(3), pp. 58–63.

[13] Wege, A., Kondak, K., Hommel, G., 2005, Mechanical design and motion control of a hand exoskeleton for rehabilitation, *IEEE International Conference on Mechatronics and Automation*, pp. 155–159.

[14] Gopura, R., Kiguchi, K., 2009, Mechanical designs of active upper-limb exoskeleton robots: State-of-the-art and design difficulties, *IEEE International Conference on Rehabilitation Robotics*, pp. 178–187.

[15] Zoss, B., Kazerooni, H., Chu A., 2006, Biomechanical design of the Berkeley lower extremity exoskeleton (BLEEX), *IEEE/ASME Transactions on Mechatronics*, **11**(2), pp. 128–138.

[16] Chu, A., Kazerooni, H., Zoss, A., 2005, On the biomimetic design of the berkeley lower extremity exoskeleton (BLEEX), *Proceedings of the 2005 IEEE International Conference on Robotics and Automation*, pp. 4345–4352.

[17] Yokokohji, Y., Yoshikawa, T., 1994, Bilateral control of master-slave manipulators for ideal kinesthetic coupling-formulation and experiment. *IEEE Transactions on Robotics and Automation*, **10**(5), pp. 605–620.

[18] Goertz, R. C., 1964, November, Manipulator systems development at ANL. *12th Conference on remote systems Technology Proceedings of the, ANS*, pp. 117–136.

[19] Tsuji, T., Tanaka, Y., 2005, Tracking control properties of human-robotic systems based on impedance control. *Systems,*

IEEE Transactions on Man and Cybernetics, Part A: Systems and Humans, **35**(4), pp. 523–535.

[20] Ruiz, A. F., Rocon, E., Raya, R., Pons, J. L., 2008, May, Coupled control of human-exoskeleton systems: An adaptative process. *2008 Conference on Human System Interactions,* pp. 242–246.

[21] Chu, A., Kazerooni, H., Zoss, A., 2005, April, On the biomimetic design of the berkeley lower extremity exoskeleton, BLEEX, *2005 IEEE International Conference on Robotics and Automation, 2005. ICRA 2005, Proceedings of the,* pp. 4345–4352.

[22] Bien, Z., Song, W. K., 2003, Blend of soft computing techniques for effective human–machine interaction in service robotic systems, *Fuzzy Sets and Systems,* **134**(1), pp. 5–25.

[23] Alkan, A., Günay, M., 2012, Identification of EMG signals using discriminant analysis and SVM classifier, *Expert Systems with Applications,* **39**(1), pp. 44–47.

[24] Nicolelis, M. A., 2001, Actions from thoughts, *Nature,* **409**(6818), pp. 403–407.

[25] Lenzi, T., De Rossi, S. M. M., Vitiello, N., Carrozza, M. C., 2012, Intention-based EMG control for powered exoskeletons, *Biomedical Engineering, IEEE Transactions on,* **59**(8), pp. 2180–2190.

[26] Pratt, G. A., Williamson, M. M., 1995, August, Series elastic actuators, *1995 IEEE/RSJ International Conference on Intelligent Robots and Systems 95.' Human Robot Interaction and Cooperative Robots', Proceedings,* **1**, pp.399–406.

[27] Kaminaga, H., Tanaka, H., Nakamura, Y., 2011, Mechanism and Control of Knee Power Augmenting Device with Backdrivable Electro-Hydrostatic Actuator, *13th World Congress in Mechanism and Machine Science,* **12**, p. 534.

[28] Kato, I., 1973, Development of WABOT 1, *Biomechanism,* **2**, pp. 173–214.

[31] Sakagami, Y., Watanabe, R., Aoyama, C., Matsunaga, S., Higaki, N., Fujimura, K., 2002, The intelligent ASIMO: System overview and integration, *IEEE/RSJ International Conference on Intelligent Robots and Systems,* **3**, pp. 2478–2483.

[32] Hirukawa, H., 2007, Walking biped humanoids that perform manual labour, *Philosophical Transactions of the Royal Society A: Mathematical, Physical and Engineering Sciences,* **365**(1850), pp. 65–77.

[33] Hirose, M., Ogawa, K., 2007, Honda humanoid robots development. *Philosophical Transactions of the Royal Society A: Mathematical, Physical and Engineering Sciences*, **365**(1850), pp. 11–19.
[34] Ogura, Y., Shimomura, K., Kondo, H., Morishima, A., Okubo, T., Momoki, S., Takanishi, A., 2006, October, Human-like walking with knee stretched, heel-contact and toe-off motion by a humanoid robot, *2006 IEEE/RSJ International Conference on Intelligent Robots and Systems*, pp. 3976–3981.
[35] Kaneko, K., Kanehiro, F., Morisawa, M., Akachi, K., Miyamori, G., Hayashi, A., Kanehira, N., 2011, September, Humanoid robot hrp-4-humanoid robotics platform with lightweight and slim body, *2011 IEEE/RSJ International Conference on Intelligent Robots and Systems*, pp. 4400–4407.
[36] Ito, Y., Urata, J., Nakanishi, Y., Okada, K., Inaba, M., 2011, October, Design and development of shock absorbable tendon driven ankles for high-powered humanoids, *2011 IEEE Workshop on Advanced Robotics and its Social Impacts*, pp. 58–61.
[37] Urata, J., Nshiwaki, K., Nakanishi, Y., Okada, K., Kagami, S., Inaba, M., 2012, October, Online walking pattern generation for push recovery and minimum delay to commanded change of direction and speed, *2012 IEEE/RSJ International Conference on Intelligent Robots and Systems*, pp. 3411–3416.
[38] Huang, Q., Li, K., Wang, T., 2002, Control and mechanical design of humanoid robot BHR-01, *Proceeding of The Third IARP International Workshop on Humanoid and Human Friendly Robotics*, pp. 10–13.
[39] Zhang, L., Huang, Q., Lu, Y., Xiao, T., Yang, J., Keerio, M., 2006, October, A Visual Tele-operation System for the Humanoid Robot BHR-02, *2006 IEEE/RSJ International Conference on Intelligent Robots and Systems*, pp. 1110–1114.
[40] Xu, W., Huang, Q., Li, J., Yu, Z., Chen, X., Xu, Q., 2011, May, An improved ZMP trajectory design for the biped robot BHR, *2011 IEEE International Conference on Robotics and Automation*, pp. 569–574.
[41] Favot, V., Buschmann, T., Schwienbacher, M., Ewald, A., Ulbrich, H., 2012, November, The sensor-controller network of the humanoid robot LOLA, *2012 12th IEEE-RAS International Conference on Humanoid Robots*, pp. 805–810.

[42] Tajima, R., Honda, D., Suga, K., 2009, May, Fast running experiments involving a humanoid robot, *2009 ICRA'09 IEEE International Conference on Robotics and Automation*, pp. 1571–1576.
[43] Kim, M. S., Kim, M. S., Park, S., Oh, J. H., 2008, December, Realization of stretch-legged walking of the humanoid robot, *2008 8th IEEE-RAS International Conference on Humanoid Robots*, pp. 118–124.
[44] Cho, B. K., Oh, J. H., 2009, July, Running pattern generation of humanoid biped in the three-dimensional space and its realization, *2009 IEEE/ASME International Conference on Advanced Intelligent Mechatronics*, pp. 142–149.
[45] Boston Dynamics, PETMAN[EB/OL], 2012-4-10, http://www.bostondynamics, com/robot_ Atlas.html.
[46] Hyon, S. H., Cheng, G., 2006, Gravity compensation and full-body balancing for humanoid robots, *Humanoid Robots, 2006 6th IEEE-RAS International Conference on*, pp. 214–221.
[47] Stephens, B. J., Atkeson, C. G., 2010, Dynamic balance force control for compliant humanoid robots, *2010 IEEE/RSJ International Conference on Intelligent Robots and Systems*, pp. 1248–1255.
[48] Pratt, J., Krupp, B., 2008, April, Design of a bipedal walking robot, *SPIE Defense and Security Symposium*, pp. 69621F–69621F.
[49] Pratt, J., Koolen, T., De Boer, T., Rebula, J., Cotton, S., Carff, J., Neuhaus, 2012, Capturability-based analysis and control of legged locomotion, Part 2: Application to M2V2, a lower-body humanoid, *The International Journal of Robotics Research*, **31**(10), pp. 1117–1133.
[50] Pratt, J., Pratt, G., 1998, May, Intuitive control of a planar bipedal walking robot, *1998 IEEE International Conference on Robotics and Automation*, **3**, pp. 2014–2021.
[51] Vermeulen, J., Verrelst, B., Vanderborght, B., Lefeber, D., Guillaume, P., 2006, Trajectory Planning for the Walking Biped "Lucy", *The International Journal of Robotics Research*, **25**(9), pp. 867–887.
[52] Vanderborght, B., 2010, Dynamic stabilisation of the biped Lucy powered by actuators with controllable stiffness, **63**.
[53] Vanderborght, B., Van Ham, R., Lefeber, D., Sugar, T. G., Hollander, K. W., 2009, Comparison of mechanical design and energy consumption of adaptable, passive-compliant actuators, *The International Journal of Robotics Research*, **28**(1), pp. 90–103.

[54] Asano, Y., Mizoguchi, H., Kozuki, T., Motegi, Y., Osada, M., Urata, J., Aba, M., 2012, Lower thigh design of detailed musculoskeletal humanoid "Kenshiro", *IEEE/RSJ International Conference on Intelligent Robots and Systems*, pp. 4367–4372.
[55] Nakanishi, Y., Asano, Y., Kozuki, T., Mizoguchi, H., Motegi, Y., Osada, M., Inaba, M., 2012, November, Design concept of detail musculoskeletal humanoid "Kenshiro"-Toward a real human body musculoskeletal simulator, *12th IEEE-RAS International Conference on Humanoid Ro*bots, pp. 1–6.
[56] Ott, C., Baumgartner, C., Mayr, J., Fuchs, M., Burger, R., Lee, D., ... Hirzinger, G., 2010, December, Development of a biped robot with torque controlled joints, *10th IEEE-RAS International Conference on In Humanoid Robots*, pp. 167–173.
[57] Kuo, A. D., 2002, Energetics of actively powered locomotion using the simplest walking model, *Journal of Biomechanical Engineering*, **124**(1), pp. 113–120.
[58] Asano, F., Yamakita, M., Kamamichi, N., Luo, Z. W., 2004, A novel gait generation for biped walking robots based on mechanical energy constraint, *IEEE Transactions on Robotics and Automation*, **20**(3), pp. 565–573.
[59] Wisse, M., 2005, Three additions to passive dynamic walking: actuation, an upper body, and 3D stability, *International Journal of Humanoid Robotics*, **2**(4), pp. 459–478.
[60] Wisse, M., Feliksdal, G., Van Frankkenhuyzen, J., Moyer, B., 2007, Passive-based walking robot, *Robotics Automation Magazine, IEEE*, **14**(2), pp. 52–62.
[61] Hobbelen, D., de Boer, T., Wisse, M., 2008, September, System overview of bipedal robots flame and tulip: Tailor-made for limit cycle walking, *IEEE/RSJ International Conference on Intelligent Robots and Systems*, pp. 2486–2491.
[62] Bhounsule, P., Ruina, A., 2009, Cornell ranger: energy-optimal control. *Dynamic walking*.
[63] Tedrake, R., Zhang, T. W., Fong, M. F., Seung, H. S., 2004, April, Actuating a simple 3D passive dynamic walker, *2004 IEEE International Conference on Robotics and Automation, 2004, Proceedings*, Vol. **5**, pp. 4656–4661.
[64] Tedrake, R., Zhang, T. W., Seung, H. S., 2004, Stochastic policy gradient reinforcement learning on a simple 3D biped, *2004*

IEEE/RSJ International Conference on Intelligent Robots and Systems, Vol. **3**, pp. 2849–2854.

[65] Sakagami, Y., Watanabe, R., Aoyama, C., Matsunaga, S., Higaki, N., Fujimura, K., 2002, The intelligent ASIMO: System overview and integration, *2002, IEEE/RSJ International Conference on Intelligent Robots and Systems*, **3**, pp. 2478–2483.

[66] Kaneko, K., Kajita, S., Kanehiro, F., Yokoi, K., Fujiwara, K., Hirukawa, H., Isozumi, T., 2002, Design of advanced leg module for humanoid robotics project of METI, *IEEE International Conference on Robotics and Automation*, **1**, pp. 38–45.

[67] Lohmeier, S., Buschmann, T., Ulbrich, H., 2009, May, Humanoid robot LOLA, *IEEE International Conference on Robotics and Automation*, pp. 775–780.

[68] Miura, K., Morisawa, M., Kanehiro, F., Kajita, S., Kaneko, K., Yokoi, K., 2011, September, Human-like walking with toe supporting for humanoids, *2011 IEEE/RSJ International Conference on Intelligent Robots and Systems*, pp. 4428–4435.

[69] Hashimoto, K., Motohashi, H., Takashima, T., Lim, H. O., Takanishi, A., 2013, May, Shoes-wearable foot mechanism mimicking characteristics of human's foot arch and skin, *Robotics and Automation, ICRA, 2013 IEEE International Conference on*, pp. 686–691.

[70] Omer, A. M. M., Ogura, Y., Kondo, H., Morishima, A., Carbone, G., Ceccarelli, M., Takanishi, A., 2005, Development of a humanoid robot having 2-DOF waist and 2-DOF trunk, *2005 5th IEEE-RAS International Conference on Humanoid Robots*, pp. 333–338.

[71] Perry, J., Davids, J. R., 1992, Gait analysis: normal and pathological function, *Journal of Pediatric Orthopaedics*, **12**(6), p. 815.

[72] Zhang, S., Huang, Q., Wang, H., Xu, W., Ma, G., Liu, Y., Yu, Z., 2013, The Mechanism of Yaw Torque Compensation in the Human and Motion Design for Humanoid Robots, *International Journal of Advanced Robotic Systems*, **10**.

[73] Ulbrich, H., Buschmann, T., Lohmeier, S., 2006, Development of the humanoid robot LOLA, *Applied Mechanics and Materials*, **5**, pp. 529–540.

[74] Vukobratovic, M., Juricic, D., 1969, Contribution to the synthesis of biped gait, *IEEE Transactions on Biomedical Engineering*, **1**, pp. 1–6.

[75] Kajita, S., Kanehiro, F., Kaneko, K., Fujiwara, K., Yokoi, K., Hirukawa, H., 2002, A realtime pattern generator for biped walking, *IEEE International Conference on Robotics and Automation*, **1**, pp. 31–37.

[76] Lim, H. O., Takanishi, A., 2001, Realization of continuous biped walking, *2001 IEEE International Conference on Systems, Man, and Cybernetics*, **3**, pp. 1630–1635.

[77] Choi, Y., Kim, D., You, B. J., 2006, May, On the walking control for humanoid robot based on the kinematic resolution of com jacobian with embedded motion, *2006 IEEE International Conference on Robotics and Automation*, pp. 2655–2660.

[78] Wieber, P. B., 2006, Trajectory free linear model predictive control for stable walking in the presence of strong perturbations, *2006 6th IEEE-RAS International Conference on Humanoid Robots*, pp. 137–142.

[79] Takenaka, T., Matsumoto, T., Yoshiike, T., Hasegawa, T., Shirokura, S., Kaneko, H., Orita, A., 2009, Real time motion generation and control for biped robot-4 th report: tegrated balance control, *IEEE/RSJ International Conference on Intelligent Robots and Systems*, pp. 1601–1608.

[80] Sugihara, T., 2009, May, Standing stabilizability and stepping maneuver in planar bipedalism based on the best COM-ZMP regulator, *IEEE International Conference on Robotics and Automation*, pp. 1966–1971.

[81] Kajita, S., Morisawa, M., Miura, K., Nakaoka, S., Harada, K., Kaneko, K., Yokoi, K., 2010, Biped walking stabilization based on linear inverted pendulum tracking, *2010 IEEE/RSJ International Conference on Intelligent Robots and Systems*, pp. 4489–4496.

[82] Pratt, J., Carff, J., Drakunov, S., Goswami, A., 2006, Capture point: A step toward humanoid push recovery, *2006 6th IEEE-RAS International Conference on Humanoid Robots*, pp. 200–207.

[83] Englsberger, J., Ott, C., Roa, M. A., Albu-Schaffer, A., Hirzinger, G., 2011, September, Bipedal walking control based on capture point dynamics, *2011 IEEE/RSJ International Conference on Intelligent Robots and Systems*, pp. 4420–4427.

[84] Huang, Q., Nakamura, Y., 2005, Sensory reflex control for humanoid walking, *IEEE Transactions on Robotics*, **21**(5), pp. 977–984.

[85] Hirai, K., Hirose, M., Haikawa, Y., Takenaka, T., 1998, May, The development of Honda humanoid robot, *1998 IEEE International Conference on Robotics and Automation*, **2**, pp. 1321–1326.

[86] Kaneko, K., Harada, K., Kanehiro, F., Miyamori, G., Akachi, K., 2008, September, Humanoid robot HRP-3, *IEEE/RSJ International Conference on Intelligent Robots and Systems*, pp. 2471–2478.

[87] Kanehiro, F., Ishiwata, Y., Saito, H., Akachi, K., Miyamori, G., Isozumi, T., Hirukawa, H., 2006, Distributed control system of humanoid robots based on real-time ethernet, *2006 IEEE/RSJ International Conference on Intelligent Robots and Systems*, pp. 2471–2477.

[88] Kim, J. Y., Park, I. W., Lee, J., Kim, M. S., Cho, B. K., Oh, J. H., 2005, April, System design and dynamic walking of humanoid robot KHR-2, *2005 IEEE International Conference on. Robotics and Automation*, pp. 1431–1436.

[89] Kuroki, Y., Fujita, M., Ishida, T., Nagasaka, K. I., Yamaguchi, J. I., 2003, September, A small biped entertainment robot exploring attractive applications, *IEEE International Conference on Robotics and Automation, 2003, Proceedings*, **1**, pp. 471–476.

[90] Lohmeier, S., Loffler, K., Gienger, M., Ulbrich, H., Pfeiffer, F., 2004, April, Computer system and control of biped "Johnnie". In *Robotics and Automation, 2004. 2004 IEEE International Conference on Proceedings*, **4**, pp. 4222–4227.

[91] Chen, X., Huang, Q., Yu, Z., Li, J., Xu, W., Meng, F., Liu, J., 2010, August, Design and experiment of an open control system for a humanoid robot, *IEEE International Conference on Automation and Logistics*, pp. 367–372.

[92] Takano, W., Yamane, K., Nakamura, Y., 2007, Capture Database through Symbolization, Recognition and Generation of Motion Patterns, *ICRA*, pp. 3092–3097.

[93] Huang, Q., Yu, Z., Zhang, W., Duan, X., Huang, Y., Li, K., 2008, Generation of humanoid walking pattern based on human walking measurement, *2008, 8th IEEE-RAS International Conference on Humanoid Robots*, pp. 99–104.

[94] Ha, S., Han, Y., Hahn, H., 2008, August, Natural gait generation of biped robot based on captured human motion image, *IEEE*

International Conference on Multisensor Fusion and tegration for telligent Systems, pp. 522–525.

[95] Miura, K., Morisawa, M., Nakaoka, S., Kanehiro, F., Harada, K., Kaneko, K., Kajita, S., 2009, Robot motion remix based on motion capture data towards human-like locomotion of humanoid robots, *IEEE-RAS International Conference on Humanoid Robots*, pp. 596–603.

[96] Yamane, K., Hodgins, J., 2009, Simultaneous tracking and balancing of humanoid robots for imitating human motion capture data, *IEEE/RSJ International Conference on Intelligent Robots and Systems*, pp. 2510–2517.

[97] Kim, S. K., Hong, S., Kim, D., Oh, Y., You, B. J., Oh, S. R., 2010, Online footprint imitation of a humanoid robot by walking motion parameterization, *2010 IEEE/RSJ International Conference on Intelligent Robots and Systems*, pp. 2692–2697.

[98] Boutin, L., Eon, A., Zeghloul, S., Lacouture, P., 2010, An auto-adaptable algorithm to generate human-like locomotion for different humanoid robots based on Motion Capture Data, *2010 IEEE/RSJ International Conference on Intelligent Robots and Systems*, pp. 1256–1261.

[99] Hashimoto, K., Kang, H. J., Nakamura, M., Falotico, E., Lim, H. O., Takanishi, A., Berthoz, A., 2012, Realization of biped walking on soft ground with stabilization control based on gait analysis, *2012 IEEE/RSJ International Conference on Intelligent Robots and Systems*, pp. 2064–2069.

[100] Hashimoto, K., Kang, H. J., Nakamura, M., Falotico, E., Lim, H. O., Takanishi, A., Berthoz, A., 2012, Realization of biped walking on soft ground with stabilization control based on gait analysis, *2012 IEEE/RSJ International Conference on Intelligent Robots and Systems*, pp. 2064–2069.

[101] Huang, Q., Yang, J., Yu, Z., Xu, W., Li, J., Li, K., 2007, Measurement of human walking and generation of humanoid walking pattern, *IEEE International Conference on Robotics and Biomimetics*, pp. 127–132.

[102] Huang, Q., Yokoi, K., Kajita, S., Kaneko, K., Arai, H., Koyachi, N., Tanie, K., 2001, Planning walking patterns for a biped robot, *IEEE Transactions on Robotics and Automation*, **17**(3), pp. 280–289.

3. Mechanical design challenges in rescue robot prototyping

Emin Faruk Kececi, Istanbul Technical University, Turkey

Abstract: In rescue robotics, mobile robot platforms are used for search and rescue operations in the disaster zone. The design and realization of the robot from the idea to a working prototype require extensive knowledge and experience in mechanical, electrical and computer engineering. In the design and construction steps, mathematical calculations and manufacturing steps are carried out to realize the robot. Both digital and physical prototyping are explained and their differences and necessities are clarified. Later in the optimization stage risk analysis, functional conflicts, availability of materials, manufacturing capabilities and testing are considered and included in the design of a final prototype for better performance. This chapter explains the practice of rescue robot prototyping and the knowledge explained in this study can be applied to other mobile robotic fields.

3.1 Introduction

Rescue robotics is the field of mobile robotics, where the robots are designed and used for search and rescue operations after natural and man-made disasters [1, 2]. The most common features of the rescue robots is their usage, where the robot is used to find the victims under the rubble. For this purpose some researchers use flying robots to observe the disaster zone from the above [3, 4] and other type of robots are the ground vehicles [5, 6]. The ground vehicles are driven into the rubble and possible survivors are searched for inside the damaged buildings where a flying robot cannot go in.

A search and rescue robot is used mainly to find survivors in the disaster zone. The disaster can be either natural or man-made, and apart from traveling on rough or soft terrain, the robot needs to send sensory feedback to the users. This feedback can be a camera view or sound in the environment as well as gas levels or light level [7, 8].

No existing platform can qualify to be emergency response team equipment. Since there is still a need for a reliable platform design, a new robotic platform is being designed and prototyped where first of all, the locomotion mechanism is decided and the robot is built around this concept. The theoretical calculations are carried out for drivetrain and battery requirements. By using computer aided design software, the digital prototype of the robot is achieved. At the digital prototype stage, the placement of the parts is also decided and a model is created for computer aided analysis. With computer aided analysis, strength, vibration, part optimization and impact resistant simulations are run to see the design weaknesses to be improved with iterations. After the digital prototype is completed, manufacturing drawings are generated mechanical parts are manufactured and assembled. With the integration of the electronic circuits, the robot is ready to be programmed. After the embedded electronics on the robot are programmed, it is ready to move with the control signals coming from the operator or autonomously [9].

Methodology of prototyping is explained and updated to optimize the design. However, it lacks the details of manufacturing and of prototyping [10]. The existing literature on prototyping explains the general process and the cycle of optimization to improve the design and the product [11]. However, the current literature on robotics is very limited and the interaction of mechanical, electronics and software parts in the optimization process is an ongoing research. This chapter is intended to fill this gap and teach the general methodology of robot prototyping.

In this chapter, digital and physical prototyping are explained and problem solving methods in physical prototyping are discussed to help designers with their new designs. The methodology explained in this chapter can be applied not only to prototyping of ground vehicles, but also to prototyping a general mobile robot.

The following sections of this chapter explain how the digital and physical prototyping are achieved. When the digital prototyping provides a computer model of the system, main ideas can be drawn and basic analysis can be achieved. Physical prototyping involves a design process where the digital prototype data is used and parts are manufactured and assembled. Fault analysis in physical prototyping involves noticing the possible problems ahead of time and fixing them before they stop the whole prototyping process. After all, prototyping is an iterative process and product design and manufacturing stages must be updated to be successful.

3.2 Design challenges

The main mechanical challenges of rescue robotics are mobility and energy. It is very important to make a rescue robot as small as possible so that it can penetrate into the rubble easier. However, when the robot gets smaller, any obstacle becomes bigger when it is compared to its own drive system, making the mobility of the robot limited. Communication with the robot is mostly achieved with a wireless unit, where coverage becomes a fundamental issue and should be guaranteed for a successful mission. Measurement is done both internally and externally to understand the situation of the robot and the environment. Sensor fusion still poses a great challenge [12–14].

In order for the robot to move on a confined and difficult terrain, mobility becomes one of the most important features of the robot. After all, if the robot cannot move inside and on the rubble, it is simply useless. In order to increase the mobility, different designs such as wheeled, legged and tracked designs [15–17] have been achieved.

The energy requirement of the robot defines the endurance of the robot. However, simply increasing the battery size is not a solution, since both the size and the weight will be increasing. The robot can be connected to the operating unit with a wire and the energy can be transferred with this cable but the tangling of the cable poses another problem. Reliable battery research is one key to the success of mobile robots [18].

The solution to the common rescue robot mechanical design problems lies in the optimization of weight and mechanism. The platform design to increase the mobility is an ongoing research, but the adaptability to the terrain increases the mobility. Energy performance of the robot is improved by making the robot body lighter, which is explained in detail in the following sections.

When the decrease of weight of the robot is achieved, more sensors can be placed on the robot for more accurate measurements. As an internal sensor, especially the advancements on the IMU sensors with a 3 axis accelerometer, 3 axis gyroscope, 3 axis magnetometer and an integrated GPS allow the operator to understand the position, heading and the motion of the robot. External sensors are used to avoid obstacles (such as sonar and IR range sensors) and find victims in the rubble (camera and heat sensor).

3.3 Digital prototyping

The purpose of digital prototyping is to solidify the design, to have parts' drawings, to decide manufacturing methods, to decide the assembly order and to see the feasibility of the assembly, to provide communication within the project team, to decide the necessary standard components and custom made components, to decide the placements of the components and to create the bill of materials (BOM) [19].

When the design requirements of the new platform for a rescue robot are listed, the designers will have ideas about which mechanisms to use for mobility. Depending on the requirements, the robot can be wheeled or tracked, or any other way. The same requirements also define the geometrical dimensions of the robot. The dimensions are decided with consideration to the system components and their functions. For example, the specifications and the size of the camera and its placement define the front face of the robot. When the space and placement of every each part of the robot are considered and decided, the shape of the robot is finalized. Figure 3-1 shows the digital prototype of a rescue robot.

One very important point at this stage is the cabling of electronics parts, motors, batteries and sensors. Since a simple robot will have almost 100 power and signal cables in between the electronic equipment, the cables are too many to draw on the digital prototype. Moreover, the computer file becomes too big to work with. As a result, the space left for the cables can be inadequate, and even more importantly the cables can cross over each other and thus assembling becomes very time consuming.

Figure 3-1 Digital prototype of a rescue robot.

Theoretical calculations are carried out for the selection of the motor torque and speed. Since the expected weight of the robot is limited, for the maximum weight and desired speed, the drivetrain motors are calculated and chosen from the manufacturer's catalogs. The batteries are calculated for the required endurance with the voltage and ampere requirements of the motors. At the selection of the batteries the discharge rate (C) of the battery should also be taken into consideration [20].

Mechanical parts of the body of the robot are designed considering the failure modes and the strength of the parts. The important stage of the design is part optimization. By using computer aided engineering programs, the strength analysis is carried out and the parts are made lighter for weight reduction. Reducing the total weight of the robot increases the endurance [21].

After the digital prototyping is completed, a model of the robot is ready to be used for manufacturing. After a careful study of each part, design for manufacturing is achieved and manufacturing methods are decided. Assembly is one of the most important parts of the manufacturing stage in which problems occur the most. The placing and the integration of the different parts of the robot should be taken into consideration, in order to avoid any flushing of the parts. Moreover, space for the assembly equipment should be planned.

The computer aided drawing programs are developed enough to show the design in a 3 dimensional view. This ease of use allows for the team of mechanical designers, electrical engineers and programmers to work more closely and to understand each other's work better.

3.4 Physical robot prototyping

The main purpose of physical prototyping is to test the digital prototype physically. The main problem with the digital prototype is that the computer software makes assumptions and some of these assumptions are not correct or not valid. A simple example of this type of assumption is that on a tracked robot, the software cannot determine the expansion of the track under the driving force.

There are 4 reasons to prototype: communication, proof of the theory, integration and milestone [22]. During the prototyping stage many parties are involved apart from the engineers, such as the vendors, management, customers and investors. The prototype can help to explain the look and the functions – it can basically give the feeling of the product.

For example, the portability of a rescue robot is very important, since it needs to be carried by the user. When the involved parties lift and carry the physical prototype, they can really understand the importance of the weight of each piece of the robot. Figure 3-2 shows an actual prototyped rescue robot. The real prototype is also used for physical testing.

In some cases of the design process, it is not clear to the design team how the physical system will work. This occurs because of the lack of knowledge. For example, when the material of the tracks is chosen generally the friction coefficient between the track and the rubble surface is not known. With experiments this information can be obtained. In order to do the experiments, a prototype must be built.

At this initial stage, the prototype can have only the interested section, the other parts of the prototype can be represented elements. In other words, when the friction of the tracks is investigated, there is no need to build a complete robot, which would increase the cost drastically. Rather than using exact batteries during the friction tests, the robot can be driven from a power source and a dead weight can be placed inside the robot to represent the battery weight. This way the current requirement of the system can be easily observed and the place of the batteries can be changed, since there is no wire attached from

Figure 3-2 The prototyped 4 leg-tracked rescue robot.

the dead weight. Concurrent engineering allows the designers from different fields to work together and improve a product at the design stage [23].

At the more advanced stages of prototyping, the integration of the different subsystems becomes vital. In the digital prototype it is almost impossible to draw every part of the robot. Especially the connection elements and cables take too much memory space. Not modeling the connection elements, namely nuts and bolts can cause some serious problems in assembly. The bolts hit each other and sometimes there is not enough space for the tools to work inside the robot. When the cables are considered, a dual DC motor driver needs 10 signal cables and 6 power cables. When the microcontroller, battery, motor drivers, sensors and switch cables are considered, in a normal size rescue robot (10 × 50 × 30 cm – h × l × w) there is almost 4 meters of cables. The main problem is not the length of the cables, rather the number (almost 100) and the volume of the cables.

In the process of physical prototyping, design and calculations are carried out at the design process stage. In this part, the designer is more concerned about how to make the system work. Later fault analysis should follow, considering how the system might not work and necessary precautions should be studied. Especially, for the improvement of the prototype, where for a given function the design is changed, the effect of the design changes on the rest of the design must be carefully studied. The given function can be cost or total weight.

For example, to decrease the manufacturing cost, connection elements can be decreased and their places can be changed. These changes may result in a decrease in the rigidity of the structure and the new placement of the connection elements can cause some conflicts.

Weight reduction can be achieved with more efficient components, such as batteries and motors or even motor drivers. However, the placement of these new components should be carefully studied for the inside packing of the robot and the position of the center of gravity must be recalculated.

At the end of the physical prototyping, the bill of materials is created to see the necessary parts and the cost of the product. The list will also include the custom made parts and standard components to be purchased. Naturally the cost of the first prototype will be much higher, since the volume is low.

3.5 Design process

Mechanical design is achieved depending on the function of the robot. The choice of mechanism affects the mobility of the robot. Some of these mechanisms are outside of the robot, such as a flap with a track on it to increase the ability to place the track at a desired position for higher traction force. In some cases the motor is not directly connected to the driving wheel or track, where transmission elements (belts or chains) and housing elements are used. Selection and sizing of the power transmission elements and housing and bearing elements affect the overall size and weight of the structure. The system optimization parameters should be taken into consideration at the selection of the mechanism design step. Most of the time, size and weight are the most important parameters.

In mathematical modeling, kinematic and the dynamic analysis are done, where the kinematic analysis shows how the mechanisms will work and the dynamic analysis proves the necessary torque values. The dynamic equations are later used in the controller design. The forces acting on the tracks of the robot during steering are identified and the dynamical equations of the platform are found with Lagrangian method.

Depending on the desired speed and the torque, the motors are selected. At this stage voltage and current ratings of the motor should be decided. Since size and weight are very important for a rescue robot, a more efficient motor, namely a motor with a better gearbox, generally suits better. However, it should be kept in the designer's mind that better parts will be increasing the cost of the production. Especially planetary gearboxes are very efficient and compact; however, they are more expensive. If the mechanisms allows, only one directional gearbox can be selected for higher efficiency.

The mechanical manufacturing and assembly are carried out to build the prototype physically. In mechanical manufacturing, design for manufacturing becomes very important, where the designer considers the manufacturing process and capabilities at the design stage of the parts. Mechanical assembling finalizes the hardware prototyping.

The result of this stage will be a working prototype, but the performance, the extreme cases of how the robot will fail and which precautions should be taken are studied in the following section.

3.6 Fault analyses

When a prototype of a robot is built, some parts of the robot are purchased and some of the parts are designed and manufactured in house. Both the purchased parts and the manufactured parts have their own risks, such as availability, cost, adaptability and accuracy.

The risk in purchased parts is that some parts cannot be found in limited time, or the price is too high for the project budget. This limitation can cause a drop in the robot performance. For example a gearbox DC motor falls into this category, where the company can deliver the motor approximately in 8 to 10 weeks. The designer needs to continue the design with the available products on the market at the desired project time, and if absolutely necessary, purchase the parts with fast delivery time.

Another risk in purchased parts, even though the delivery time can be acceptable with the project calendar, the cost of the product can be too high. At this stage the designer has 2 choices: either to drop the performance of the robot or to ask for more funding.

The third risk of purchased components is the adaptability of the components to each other and to manufactured parts. When a motor shaft is not the same size as the pulley or the wheel to be attached, connecter hubs can be used.

The final risk of purchased versus designed and manufactured parts is the accuracy of the parts. In house manufacturing will be costly both financially and timely. If a mechanical part is not always manufactured regularly, there will be a need for additional cutting tools which will increase the unit manufacturing price. The lack of experience in manufacturing of this part will also cause time loss. If the delivery time and cost is acceptable, it is better to purchase the part rather than to manufacture it in house at this stage.

However, if the part has a strategic importance to the project and requires know-how, the manufacturing should be done in house to protect the knowledge.

3.6.1 Functional conflicts

Functional conflicts occur when the expected parts cannot be found at the design and manufacturing stages or the initial performance requirements of the robot are changed after the initial design. In the design process it is not possible to improve any feature of the robot, without causing any penalty. For example, if the speed of the robot is to be increased, stronger motors will be required which are generally bigger and thus will cause the robot to be heavier than the initial case. In return, the weight increase will require more torque from the motors. At the design stage, availability of the parts must be carefully studied.

The performance criteria of the robot must be identified at the beginning of the design stage and the important parameters should be decided. Some of the most important parameters are size, weight, endurance, mobility, speed, load capacity and water and dust resistance. The important factors depend on the prototype and the design team, where the weight might be very important for portability purposes or for higher load capacity.

For example in a rescue robot when the stair climbing ability is desired, the robot needs to be longer to touch the stairs from multiple points; however, in general, a rescue robot should be smaller in size to penetrate into rubble better. The sizing problem can be solved by the robot having flaps and opening them during stair climbing to increase the contact points. Figure 3-3 shows the prototype robot on the stairs

Figure 3-3 The length constraint on the robot to adapt to the environment.

when the flaps are fully opened. By closing the flaps, the robot length decreases allowing the robot to have a smaller turning radius.

3.6.2 Materials and manufacturing methods

The material selection and dimensions of the parts are calculated and decided together at the design stage; however, the material choices on the market are very limited. Iron, steel, aluminum and plastics are easier to find where alloys of aluminum and specific materials like titanium are not common. At the prototyping stage, the designers should choose the most common materials and even the alloys should usually be avoided both for availability and cost reasons.

The same fact applies to manufacturing processes where during the prototyping stage every manufacturing method might not available to the researchers. Moreover, for the classical manufacturing methods such as milling and turning, the accuracy and surface quality is limited. If the part requires better accuracy and surface quality, additional manufacturing methods should be considered, such as honing or lapping.

Another source of risk is that some of the manufacturing companies lack of experience either in working with some materials or implementing certain manufacturing methods. Neither the designer nor the manufacturing company wants limitations of the manufacturing method to cause the design to be impossible to be achieved.

For example, in laser cutting, the minimum diameter of a hole cannot be smaller than the thickness of the sheet metal. If the laser beam is focused to a small point, the material starts melting rather than getting cut because of the increase in the temperature at a given point before the beam is directed to that point. The manufacturing method limitations should be known and taken into consideration during the design stage.

3.6.3 Testing

Testing is the final stage of prototyping where the robot is tested for the expected performances. Also the success and the failure of the robot are learned and the experience is carried on to future robot designs. Failure modes for mechanical, electrical and software parts of the robot should be identified and tested for the designed prototype and the test environment.

For rescue robots, there is a RoboCup Rescue League standard field where the 3 zones are designed as slightly, normal and moderate earthquake damaged buildings. The robot as a final product can be tested in this type of a zone; however, the designers should test their design in a controlled environment. A military standard, MIL-STD 810 F, also exists for environmental testing of the products, where different environmental effects are described and measured in this testing. However, this military standard can be too demanding for a rescue robot where IP code (IEC 60529) can be satisfactory in regards to the measurement of water resistance and dust resistance.

At the testing stage, it is better to divide the system into pieces and test every part separately. These tests are done for two different reasons: to prove the design and to observe the performance.

At first, mechanical parts can be tested by powering them with a different power source and observing the functionality. The main causes of failure at this stage are housing and connections. When the parts are not secured correctly, they move in undesired directions and either cause unexpected friction, or collide with other parts and stop functioning. The testing shows the stability of the parts and proves the design concept and manufacturing accuracy.

At the second stage when the performance of the robot is tested, such as the impact resistance of a throwable rescue robot, the robot needs to be forced for the calculated impact. Moreover, the functionality of the mechanical parts is tested. These performance tests are designed and realized depending on the function of the robot. Analytical prototyping can give an idea about the size of the impact force and the critical cross sections, but actual testing proves the quality of the design and manufacturing. Figure 3-4 shows the rescue robot climbing on soft terrain. In this test the functionality of the track mechanism is tested, where with the increasing climbing angle, the required traction force increases to make the robot climb.

Over all, testing is very time consuming and expensive but in order to prove the robot performance it is very necessary. Moreover, the experience learned from testing is carried out into the future robot design and prototyping.

3.7 Conclusions and future directions

In the success of a rescue robot, one of the biggest constraints is the mobility of the platform. If the robot cannot reach the desired location, it cannot achieve its task. In order to increase the mobility, the weight and size of the robot become important parameters where the placements

Figure 3-4 Testing of the climbing ability on soft terrain.

of the parts play a vital role. In the design process, the calculations lead to digital and physical prototypes and later in fault analysis the most common problems which cause failure of the robot are studied and the design is updated for better performance. By considering the possible failures at the design stage, time and money can be saved in the prototyping process. The general methodology explained in this research can be applied to other mobile robots as well.

3D printing is a process where the parts are manufactured by adding operation rather than classical cutting operations. The advantage of the 3D printing is that the custom made parts can be manufactured very easily. Especially these machines work with .stl files which can be exported from the common 3D drawing programs.

However, with the current 3D printing technology mostly plastic (ABS) parts can be manufactured. Moreover, since in this process the parts are manufactured with the addition process the mechanical, the strength of the parts is lower than that of solid parts. The density of the parts is also low, since air bubbles get trapped inside the part during the manufacturing process.

In the future the 3D printing technologies will improve to cover not only plastic but also metals thus the materials used in this process will improve to provide better mechanical properties.

Next, DC motors are commonly used as a mechanical power source on mobile robots. When the weight of the robot is considered, almost 40% of the total weight of the robot is caused by the motors where the body, batteries and the electronics make up the rest of the weight.

When DC motors are compared to SMA wires, the power density of SMA wires is almost 50 times more than DC motors. This means the necessary DC motor can be replaced with a SMA wire or SMA motor and the weight of the actuator can be 50 times less, allowing the total weight of the robot to decrease almost 40%.

However, the current SMA wire technology has a cooling problem. When the electricity is passed to the wire, the wire shrinks and creates force. However, when the electricity is cut, the wire needs time to cool down and come to its original length. Since the wire density is high, especially on the SMA wire, active cooling is used to increase the response time of a SMA motor.

Advancements in SMA wire technology would allow light weight actuators and thus allow the rescue robots to be lighter.

The battery technology is also advancing with the drive from mobile electronics. Especially the hand held devices require very energy dense batteries. Lithium polymer and lithium ion batteries are almost 6 times more efficient than lead acid batteries.

The weight reduction with advanced batteries can allow the robot to be lighter or endurance of the robot to increase for the same weight. Increase in the power density of a battery will help the robot designer to focus on the functions of the robot. By having more energy on board, the robot will have more actuators to accomplish different tasks.

Manufacturing of the robot parts as well as the standard parts like actuators and batteries are improving and these new developments are allowing robot designers to create more compact and durable robots.

3.8 Acknowledgment

The author would like to thank Istanbul Technical University (ITU) for its financial support through the grant 33833.

3.9 References

[1] Casper, J., Murphy, R.R., 2003, Human-Robot Interactions During The Robot-Assisted Urban Search And Rescue Response At The World Trade Center, *IEEE Transactions On Systems Man And Cybernetics Part B-Cybernetics*, **33, no. 3**, pp. 367–385.

[2] Micire, M. J., 2008, Evolution and Field Performance of A Rescue Robot, *Journal of Field Robotics*, **25**, pp. 7–30.

[3] Sevcik, K.W., Green, W.E., Oh, P.Y., 2005, Exploring Search-And-Rescue In Near-Earth Environments For Aerial Robots, *IEEE/ASME International Conference on Advanced Intelligent Mechatronics*, pp. 699–704.

[4] Roberts, J., Walker, R., 2010, Flying Robots to the Rescue [Competitions], *IEEE Robotics & Automation Magazine*, **17, no. 1**, pp. 8–10.

[5] Michaud, F., Letourneau, D., Arsenault, M., Bergeron, Y., Cadrin, R., Gagnon, F., Legault, M.A., Millette, M., Pare, J.P., Tremblay, M.C., Lepage, P., Morin, Y., and Caron, S., 2005, Multi-Modal Locomotion Robotic Platform Using Leg-Track-Wheel Articulations, *Autonomous Robots*, **18, no. 2**, pp. 137–156.

[6] Eich, M., Grimminger, F., Kirchner, F., 2008, A Versatile Stair-Climbing Robot For Search And Rescue Applications, *IEEE International Workshop on Safety, Security and Rescue Robotics*, pp. 35–40.

[7] Nourbakhsh, I.R., Sycara, K., Koes, M., Yong, M., Lewis, M., Burion, S., 2005, Human-Robot Teaming for Search and Rescue, *IEEE Pervasive Computing*, **4, no. 1**, pp. 72–79.

[8] Aoyama, H., Himoto, A., Fuchiwaki, O., Misaki, D., Sumrall, T., 2005, Micro Hopping Robot with IR Sensor for Disaster Survivor Detection, *IEEE International Workshop on Safety, Security and Rescue Robotics*, pp. 89–94.

[9] Birk, A., Markov, S., Delchev, I., Pathak, K., 2006, Autonomous Rescue Operations on the IUB Rugbot, *IEEE International Workshop on Safety, Security, and Rescue Robotics*.

[10] Alvarez, C., Juan S., Jimenez B., Jovani A., Ramirez P., Juan F., 2011, Design Cycle Of A Robot For Learning And The Development Of Creativity In Engineering, *Dyna-Colombia*, **78, no. 170**, pp. 51–58.

[11] Warfel, T. Z., 2009, *Protoyping: A Practitioner's Guide*. Rosenfeld Media.

[12] Zorriassatine, F., Wykes, C., Parkin, R. and Gindy, N., 2003, A Survey Of Virtual Prototyping Techniques For Mechanical Product Development, *Journal of Institution of Mechanical Engineers, Part B: Engineering Manufacture*, **217, no. 4**, pp. 513–530.

[13] Guo, S., Fukuda, T., Asaka, K., 2003, A New Type Of Fish-Like Underwater Microrobot, *IEEE/ASME Transactions on Mechatronics*, **8, no. 1**, pp. 136–141.

[14] Pollack, J.B., Lipson, H., Ficici, S., Funes, P., Hornby, G., Watson, R., 2000, Evolutionary Techniques in Physical Robotics, *in Evolvable Systems: From Biology to Hardware*, J. Miller, Ed. New York: Springer-Verlag, pp. 175–186.

[15] Tunwannarux, A., Hirunyaphisutthikul, S., 2005, Design Features and Characteristics of a Rescue Robot, *IEEE International Symposium on Communications and Information Technology*, **2**, pp. 1083–1087.

[16] Mcghee, R. B., Iswandhi, G.I., 1979, Adaptive Locomotion Of A Multilegged Robot Over Rough Terrain, *IEEE Transactions on Systems, Man and Cybernetics*, **9, no. 4**, pp. 176–182.

[17] Kececi, E.F., 2009, Design and Prototype of Mobile Robots for Rescue Operations, *Robotica*, **27, no. 5**, pp. 729–737.

[18] Vincent, B., Fumihide, T., Kenji, S., 2012, Autonomous Battery Management for Mobile Robots Based on Risk and Gain Assessment, *Artificial Intelligence Review*, **37, no. 3**, pp. 217–237.

[19] Ulrich, T.K., Eppinger, S.D., 2011, *Product Design and Development*, McGraw-Hill, New York.

[20] Byoungwoo, K., Gerbrand, C., 2009, Battery Materials for Ultrafast Charging and Discharging, *Nature*, **458, no. 7235**, pp. 190–193.

[21] Lundberg, C., Christensen, H., Reinhold, R., 2007, Long-Term Study of a Portable Field Robot in Urban Terrain: Field Reports, *Journal of Field Robotics*, **24, no. 8–9**, pp. 625–650.

[22] Yang, M.C., Epstein, D.J., 2005, A Study Of Prototypes, Design Activity, And Design Outcomes, *Design Studies*, **26,** pp. 649–669.

[23] Chang, K.H., Silva, J., Bryant, I., 1999, Concurrent Design and Manufacturing for Mechanical Systems, *Concurrent Engineering Research and Applications*, **7, no. 4**, pp. 290–308.

4. Networked control for mobile robots

Xilun Ding and Fan Yang, Beijing University of Aeronautics and Astronautics, China

Abstract: As network technologies have developed, networked control became a main research focus in mobile robots, as well as in many other applications. In this chapter, the advantages and the applications of networked control mobile robots are summarized. Furthermore, the chapter presents the main problems and achievements facilitated by the interplay between networked control, communication and perception. Control over the networks has to deal with drawbacks brought by network communication, such as delays and data loss. The mobility of individual robots brings new challenges to communication and perception. Further study is still needed in the coverage and localization problems of mobile sensor networks. A brief review of our Novel Robotics System for Planetary Exploration (NOROS) and some related works, such as the study of group behaviors, optimal routing design and wireless sensor networks localization, are also presented.

4.1 Introduction

Control theory concerns influencing systems in order to achieve certain output quantities to take a desired course, and it did not become an independent scientific field until the late 1940s when N. Wiener published his classic book on control [1]. World War II and space exploration motivated the development of automatic control, and HS Tsien's "Engineering Cybernetics" [2] is considered the true beginning of modern control. As our society progressed from agricultural to mechanical, steam engine, electrical, electronic and on to the computer age, control theory and its applications experienced changes from frequency-domain design, state-space approach, and discrete event and hybrid dynamic systems, to intelligent control algorithms [3]. Without exception, every breakthrough in technology has caused a milestone change or paradigm shift in automatic control. Now that we have entered the age of a networked society, computer networks are a pervasive element of everyday life. They allow us to access information at a distance and indeed to link geographically dispersed computing resources into powerful, distributed computing platforms. The advances in networking and in miniaturization of electromechanical systems introduced the field of remote control and gave birth to networked control systems (NCS). How to apply control strategies, such as PID control, optimal control, adaptive control, robust control and intelligent control, over a network, becomes meaningful work.

The original concept of network robotics can be found in tele-robotics. Though early tele-robotics research used only a direct line to control the slave robot, at the beginning of the 1990s, The World Wide Web (WWW) motivated the first integration of robotics with networking through the creation of online robot systems [4], which can deliver motion and additional information to the operator through the Internet. Later, this technology was utilized for controlling mobile robots [5]. Embedded computers and sensors are becoming ubiquitous in homes and factories, and wireless ad hoc networks or plug-and-play wired networks are becoming increasingly commonplace. This means mobile robots could compose complex systems of distributed sensors, effector and computational resources, coordinated to perform search and reconnaissance tasks exploiting the efficiency that is inherent in parallelism. Thus a lot of mobile robot prototypes controlled by networked systems have been developed in laboratories and industry.

For networked robots, which are defined as multiple robots coordinated and cooperated together by networked communication to accomplish a specified task, the communication network plays a key role in the system [6]. The application of the network is a liberator; it allows us to do things that were not previously possible with robots. Yet there are still many challenges to overcome. The key problems derive from its interdisciplinary nature. In fact, the study requires deep knowledge of automatic control, perception and communication, and the ability to capture the interplay between these disciplines. This chapter presents an overview of recent networked control mobile robots research challenges, and the main achievements regarding the key problems mentioned above, including our related work on the NOROS robots system.

The remainder of this chapter is organized as follows: The main applications of networked control mobile robots are presented in Section 4.2. Section 4.3 introduces our novel robotics system for planetary exploration. The main advantages and challenges of networked control mobile robots are introduced briefly in Section 4.4. The main problems and achievements facilitated by the interplay between networked control, communication and perception are discussed in Sections 4.5–4.7.

4.2 Applications of networked control mobile robots

Due to recent high-tech and low-cost advances in microelectronics and sensors, such as low-power CCD cameras, small microprocessors with expanded capabilities, and autonomous navigation systems using global positioning systems, a system of mobile robots, embedded computers, actuators, and sensors has tremendous potential in reconnaissance, surveillance and physical security, particularly locating and identifying hazardous targets. The coordination of multiple mobile robots can be effective and efficient in the above applications through information sharing and the formation of distributed communication networks. Multiple robots controlled to cooperatively achieve an objective have the potential to be more effective than a collection of independent robots. The significance and potential impact of networked robots is apparent from the following examples.

The manufacturing industry has always relied on integration between sensors, actuators, material-handling equipment and robots. Many factories utilize Automated Guided Vehicle (AGV) for unmanned transportation. A representative method for guiding the vehicles is using special tapes as a guide-rail [7]. In this case, it is difficult to cope with frequent structural changes in the factory. All of the guide tapes need to be re-stuck on the floor every time. Today, companies are finding it easier to reconfigure existing infrastructure by networking new robots and sensors with existing robots via wireless networks and ubiquitous computing. For example, JH Lee et al. [8, 9] implement distributed intelligent sensor network systems for AGV in factories. The environment is monitored by distributed sensor devices which connect with other distributed sensor devices and robots throughout the network (see Figure 4-1). Though the mobile robots are not equipped with sensors for self-positioning and control, they are able to accomplish tasks simply by following orders from the sensor devices in the networked environment.

In the defense industry, the US military is engaged in the large Future Combat Systems (FCS) initiative to develop network-centric approaches to deploying autonomous vehicles. The FCS could be a distributed network centric system with all of the functionality necessary to be successful on the modern battlefield, distributed among multiple

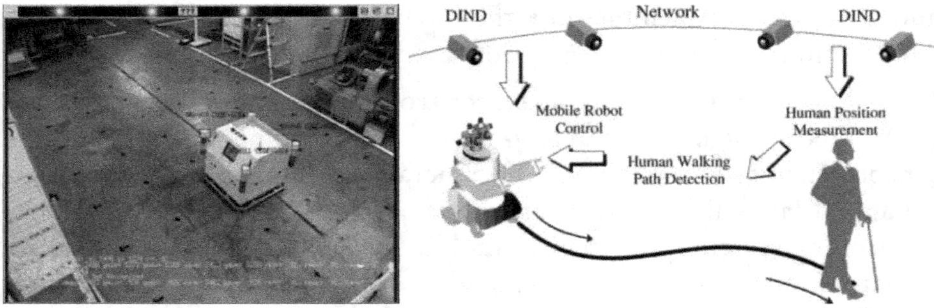

Figure 4-1 (a) Intelligent sensor networks guided AGV [9] (b) Concept of ubiquitous environments. Source: [8] (with permission).

vehicle elements whose capabilities sum to the capabilities necessary for victory in all forms of combat. The Army is accelerating the research and development of this system, and anticipates equipping its first unit with the FCS as soon as technology is available. A great deal of effort was made collaboratively between the General Robotics, Automation, Sensing & Perception (GRASP) Laboratory, the Georgia Tech Mobile Robot Laboratory and the University of Southern California's (USC) Robotic Embedded Systems Laboratory, which is funded by DARPA. They developed an autonomous, adaptive robot network capable of executing a wide range of tasks within an urban environment. It can explore urban environments and the interiors of buildings to detect and track intruders, and transmit all of the above information to a remote operator [10, 11]. Robots map an environment and deploy themselves to form a sensor network to detect intruders (see Figure 4-2).

There are already many commercial products with prohibitive costs to cope with, where large numbers of robots are needed for applications. For example, the MARON robot developed by Fujitsu lets a human user dial up their robot and instruct it to conduct simple tasks, including sending pictures back to the user via cellular phone. The K-Team Corporation is a Swiss company that develops, manufactures and markets high-quality mobile robots for use in advanced education and research. The Khepera and Koala robots they developed are a standard for academic research now, while the K-Junior robot is designed for teachers and hobbyists [12]. The iRobot Swarm [13] is comprised of over 100 SwarmBots™ (see Figure 4-3), 16 charging stations and navigational

Figure 4-2 (a) Future combat systems (b) robot network for urban environment. Source: [11] (with permission).

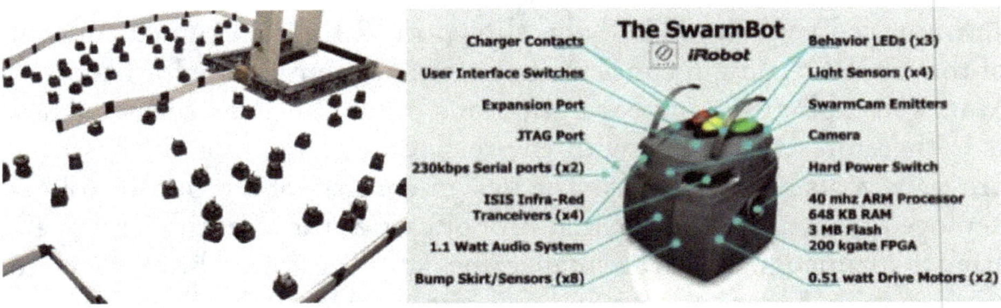

Figure 4-3 (a) iRobot Swarm; (b) SwarmBots. Source: [11] (with permission).

beacons. Each robot belonging to the swarm could easily be replaced, as they would be understandably much cheaper, which has a suite of sensors, communications hardware, and human interface devices. Indeed the Network Robot Forum [14] is already setting standards for how stationary sensors and actuators can interact with other robots in domestic and commercial settings.

Environmental monitoring is a key application for networked robots, where one can exploit the mobility and communication abilities of the robotic infrastructure for observation and data collection at unprecedented scales, in various aspects of ecological monitoring. This is significant for environmental regulatory policies (e.g., clean air and water legislation), as well as enabling new scientific discovery. Examples include systems built for aquatic monitoring [15], terrestrial monitoring [16] and subsoil monitoring. Other applications for networked robotic systems include surveillance of indoor environments [17] and support for first responders in a search and rescue operation [18].

There are also many other applications, such as the health care industry, field robotics and surveillance, and a lot of research projects are focused on group behaviors and the collective intelligence of networked mobile robots, the aim of which is improved efficiency, extra work ability, and robust fault-tolerance ability. For example, a recent multi-university US project addressed the development of networked mobile robots for swarming behaviors [19].

4.3 Introduction of NOROS robots

The Novel Robotics System for Planetary Exploration (NOROS) was developed collaboratively by Beijing University of Aeronautics and Astronautics (BUAA) and Politecnico di Milano (POLIMI), which is an improved version of "Ladyfly" shown in [20]. The original concept of the NOROS design consists of a robotic system usable for Moon or Mars soil exploration in preparation of a human mission to the planet. It is therefore compatible with the technologies relevant to the space program goals and scenarios. The design consists of a group of robots deployed from a lander. The main robot carries the scientific and operating instruments, while the smaller ones scan the area to be explored to create a map. Mapping data and soil features are then collected by the main robot and passed to the lander, then relayed to the Earth. The robots will have wheels for fast motion on flat surfaces and robotic legs to overtake obstacles, and reliable positioning for use of instruments and tools. A novel polygonal structure of the frame will be adaptable to rough terrain and to the required motion to the robot [21] (see Figure 4-4).

Then we developed the first hexapod robot prototype named NOROS-I [22, 23]. The robot is radial symmetric with hemispheric shell (see Figure 4-5), and since the legs are located symmetrically, the

Figure 4-4 Conceptual model of NOROS robot.

Figure 4-5 Prototype of NOROS-I.

angle between adjacent legs is 60°. There are three joints in each leg, namely hip, thigh and calf joint, all of which are driven by Maxon DC motors through the bevel bear. The range of joint angles of the leg is limited in (−90°, 90°), (−90°, 150°), (−90°, 150°), respectively. There are three parts inside of the robot body: the bottom part accommodates the battery and motors, the middle part contains the hardware of control system such as the CPU board, motion drive, control cards and video capture card, etc. A binocular camera that can be stretched out of the shell for environmental detection is placed at the top. The robots can walk in various gaits, and are robust to different terrain.

Meanwhile, we developed a flotilla of small robots named NOROS-II (see Figure 4-6(a)), which look like smaller versions of NOROS-I. The objective was to design a group of cheap and portable robots which are able to explore a given area rapidly, and create a global map cooperatively. Even if some robots are broken accidentally, the remaining robots of the NOROS system are still able to accomplish the task. The NOROS-II robot has a width of 980 mm and is 120 mm in height, with six legs distributed axisymmetrically around the body, which provide omni-directional mobility. Each leg consists of three revolute joints and a driving wheel. The hybrid locomotion system allows motion by legs to cross a highly soft or rough terrain, and can obtain a 3 km/hour maximum speed by moving on wheels on hard flat ground. The weight is about 4 kg, including all mechanical components, control system and lithium batteries. Different sensors are equipped for various tasks. The contact of the supporting foot can be measured by using a 1 DOF force

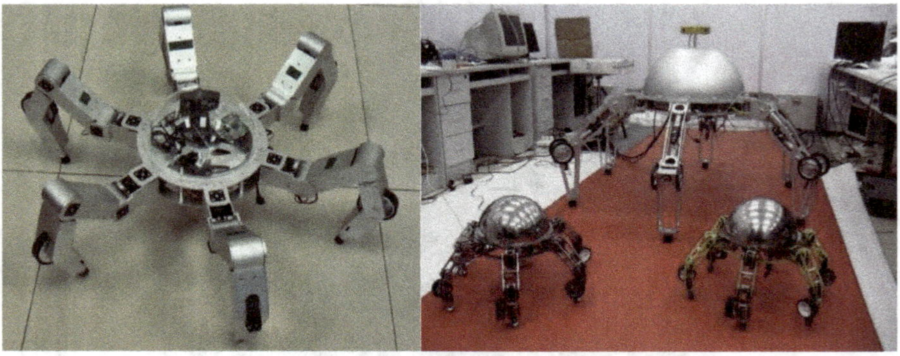

Figure 4-6 (a) Prototype of NOROS-II; (b) Groups of NOROS robots system.

sensor. A small camera and six infrared sensors are used to detect obstacles, remote control commands and the video signal are transmitted by wireless LAN, which enable us to make the robot perform various tasks at hazardous and narrow places.

We also developed a robotic team, which consists of one big robot, two small robots, the CSS and the coordinated wireless control system (see Figure 4-6(b)). A lot of experiments were carried out, such as route optimization, sensor networks, localization and mapping. The results show the wireless networked robots mapping the search area and finding features very fast in the detection area. In case one of the explorers encounters a problem, the mission can be carried on with all other robots, giving time to solve the problem while the mission goes on. The system is very robust to the application of planetary exploration.

4.4 Advantages and challenges

Mobile robots controlled by NCS share data efficiently, and it's easy to fuse the global information to take intelligent decisions over a large physical space. They eliminate unnecessary wiring, as robot architectures and systems can be liberated from the constraints of fixed wiring. It is easy to add more sensors, actuators and controllers with very little cost and without heavy structural changes to the whole system. They connect cyber space to physical space, making task execution from a distance easily accessible (a form of tele-operation). Most importantly, they can perform tasks that individual robots cannot perform [24], as ants are able to cooperatively manipulate and transport objects, often in large groups. Networked robots can potentially communicate and cooperate with each other, and even though individual robots may not be sophisticated, it is possible for networked robots to provide a range of intelligent behaviors that are beyond the scope of intelligent robots.

While there are many successful embodiments of networked robots with such applications as space exploration, disaster rescue, health care, and as museum guides [25], there are still many challenges to overcome. The key problems are derived from its interdisciplinary nature. In fact, the study requires deep knowledge in automatic control, perception and communication, and to capture the interplay between these disciplines. How to coordinate multiple mobile robots through the networks to accomplish the task? How should the information be perceived, aggregated, and distributed? These are all basic questions derived from the intersection of control theory, perception and networked communication (see Figure 4-7).

There are many challenges in the control of mobile robots caused by the characteristic features of the communication method adopted, such as the delays, unpredictable data loss, quantization, etc. A lot of contributions were devoted to solve these problems. Another main constraint

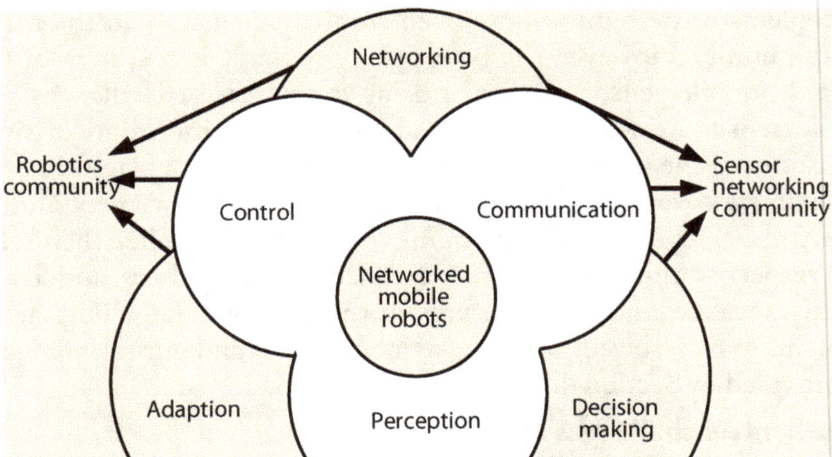

Figure 4-7 The challenges at the intersection of control, perception and communication. Source: [6] (with permission).

worth noticing is that robot networks are dynamic, unlike networks of sensors, computers or machines which might be networked together in a fixed topology, and the robot's behavior also changes as the topology changes. It is very difficult to predict the performance of such dynamic robot networks, yet it is the analytical problem that designers must solve before deploying the network. New mathematical tools are urgently needed to control individual behavior to achieve a specified aggregate motion and shape of the groups (more details are discussed in Section 4.5).

The communication subjects are mainly concerned with providing a certain level of quality of service (QOS), while achieving efficient and fair utilization of network resources. The major problems of interest are calling admission, routing, flow control and various other resource allocation problems. The mobility of an individual robot brings new challenges. It should be possible to synthesize distributed controllers to move agents to attain desired network characteristics, such as maximizing network coverage while ensuring that nodes retain line of sight with one another, or maximizing the lifetime of networks. Some basic algorithmic problems and our related works are provided in Section 4.6.

Problems of perception have been studied extensively in the robotics community. However, the perception problems in a system of networked, mobile sensor platforms bring a new set of challenges. The mobile sensor networks have the ability to reposition and organize themselves in the network. Where should the nodes be placed in order to maximize coverage? How to balance the costs of movement and the quality of perception? Robots can move in order to localize themselves with respect to their neighbors, to localize their neighbors, and also to identify, localize, and track features in the environment. These are all problems to be resolved. Several pertinent results and our related works are provided in Section 4.7.

4.5 Control challenges and achievements

4.5.1 Overview of NCS

Currently control engineers have to cope with many new problems arising from networked systems while designing complex mobile robot control systems. The advantages of controlling mobile robots over networks have been mentioned above; however there are also some challenges caused by the limitation of communication networks, such as variable delays and unpredictable data loss. Another main constraint is the changing topology of the mobile robot networks. Traditionally, models of group behavior have been built on continuous models of individual dynamics, including models of control and sensing with a fixed set of neighbors. No powerful mathematical tools have been developed to solve the changing topology problems, and further studies are still needed on the control of individual behavior to achieve a specified aggregate motion and shape of groups.

Generally speaking, the two major types of control systems that utilize communication networks are (1) shared-network control systems and (2) remote control systems [3], and the robots controlled by these systems can be called tele-operated robots and autonomous robots, respectively. The autonomous robots, using shared-network resources to transfer measurements from sensors to controllers, and control signals from controllers to actuators, can greatly reduce the complexity of connections. On the other hand, a tele-operation control system can be thought of as a system controlled by human supervisors, who send commands and receive feedback via the network. Such a system can prevent people from hazardous environments.

There are two general NCS configurations. The first is a hierarchical structure which consists of a main controller and several remote closed-loop systems (see Figure 4-8). Each of the subsystems contains a sensor, an actuator, and a controller by itself, and the main controller computes and sends the reference signal in a frame or a packet via a network to the remote system. The remote system then processes the reference signal to perform local closed-loop control, and returns to the sensor measurement to the main controller for networked closed-loop control.

The second approach of networked control is the direct structure (see Figure 4-9). This structure has a sensor and an actuator of a control loop connected directly to a network. The control signal is encapsulated in a frame or a packet and sent to the plant via the network. The plant

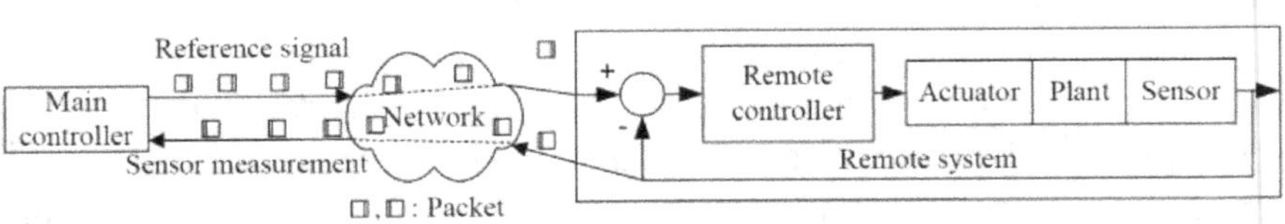

Figure 4-8 NCS in the hierarchical structure. Source: [26] (with permission).

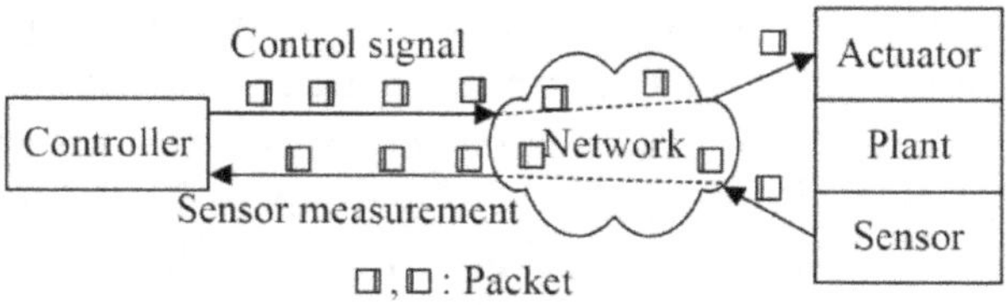

Figure 4-9 NCS in the direct structure. Source: [26] (with permission).

then returns the system output to the controller by putting the sensor measurement into a frame or a packet as well. Both the hierarchical and direct structures have their own pros and cons. Many networked control systems are a hybrid of the two structures.

4.5.2 Major accomplishments

Since an NCS operates over a network, data transfers between the controller and the remote system will suffer the drawbacks brought on by the network, such as time delays, packet loss and finite capacity. A fairly large amount of literature has been devoted to overcoming these problems. A main consideration is the delay caused by communication networks; the time to read a sensor measurement and to send a control signal to the mobile depends on network characteristics, such as delays and jitter [26]. Mainstream research has been focused on the design of control/observation algorithms that can stand the delay variations, and their unpredictability. Fattouh et al. [27], proposed a $H\infty$ method to cope with the constant time delays. For the more realistic case of variable delays, Witrant et al. [28] generalized the predictor techniques, by assuming that a dynamical ordinary differential equation model of the delay is available. In other cases, a possible solution consists of introducing an input buffer. This technique makes the delay constant, at the price of lower performance [29]. Another approach consists of combining robust control theory and network calculus theory. By estimating a bound of the end-to-end delays, Vatanski [30] proposed a robust control method, and extension of this work has been done [31] which reduces the effect of the network on the robust controller by optimizing the quality of net service. These accomplishments, considering the effect of data-loss on system stability and performance, can be roughly categorized into three types, based on the resulting closed-loop systems: switched systems, asynchronous dynamical systems and jump linear systems with Markov chains. Other works which try to overcome the drawbacks brought on by network communication are not mentioned due to the scope limitation of this chapter.

Another main approach is to deal with the coordination of multi-robots system. Groups of autonomous mobile robots with computing, communication and perception capabilities are expected to become economically feasible and perform a variety of spatially distributed sensing tasks. An interesting work is considering the problem of task partitioning among members of a team of cooperating agents in response to the discovery of new tasks or potential failures of certain agents. There is a rich literature in parallel and distributed algorithms for solving assignment problems [32, 33]. Nature provides proof of the concept that simple animals can execute simple behaviors to enable complex group behaviors, such as hunting, constructing nests and survival etc. It is still unclear how to design local laws to yield a prescribed global behavior. The picture is clearer in some specific and simple cases, such as the distributed averaging problem. But even in this simple example there are still many things to be understood. It is not yet clear what network topologies yield better performance in term of robustness to failure, flexibility, reliability and adaptability. We need a design methodology for solving the inverse problem in navigation – behaviors for controlling individuals to achieve a specified aggregate motion and shape of the group. Some experiments on group behaviors have been done using our NOROS robot groups; an example is building a team formation with two small robots to divide and simplify the exploration task. The robots converge at the top of the slope, exchange information and prepare for further exploration. The big robot fuses and analyzes the information gathered by the small robots, and makes strategies that allow the lighter robot to climb the slope by walking and the other by wheeling (Figure 4-10). However, these experiments are still in the framework of conceptual attempts to study group behaviors.

4.6 Communication challenges and achievements

4.6.1 Key problems

A communication network is central to the functioning of networked mobile robots. Multi-hop wireless networks, such as radio, ad hoc and sensor networks, are networks in which communication between two nodes may go through multiple consecutive wireless links. Unlike a

Figure 4-10 Experiments of group mapping and detection.

static network, mobile robots with transmitters and receivers can move toward each other to facilitate communication and adaptively maintain a communication network. In such an environment, due to the limited radio range of the wireless link, it may be necessary for one node to enlist the aid of other nodes in forwarding data to a destination node not within the radio transmission range of the source. Thus, each node operates not only as a host but also as a router. The movement gives us an opportunity to create desired network topologies under a suitable model of network communication, the so-called mobility-based topology control problem. It also provides an opportunity to address the information routing problem in disconnected networks by turning the robots into relay nodes; the key idea here is to enable the robot holding a current message to an unavailable destination to modify its trajectory in order to relay a message. However, there are still many challenges to resolve. How can we control the motion to ensure that all agents can talk to each other? How does a piece of information propagate through the network, and what can we say about when and where that piece of information will be heard? The movement of robots in a network of robots and sensors may cause network partitioning when nodes go out of range. Wireless networks are especially interesting from a resource control point of view. It should be possible to synthesize distributed controllers to move agents to attain desired network characteristics, such as maximizing network coverage while ensuring that nodes retain line of sight with one another, or to maximize the lifetime of ad hoc wireless networks.

4.6.2 Major accomplishments

It is important that the robots cover the space while remaining within communication range [34]. A lot of achievements have been made in this subject. Ye et al. [35], proposed a probing environment and adaptive sleeping protocol (PEAS) that can build long-lived resilient sensor networks using a large number of small sensors, which was one of the first attempts to address communication connectivity and sensing coverage simultaneously using heuristic algorithms. Wang et al. [36], proposed a new coverage configuration protocol (CCP) that can provide different degrees of coverage as requested, to provide guaranteed coverage and connectivity configurations, through both geometric analysis and extensive simulations. Zhang and Hou [37] proved that, given a fixed node density in a finite (but reasonably large) region, the upper bounds of life time when only α-portion of the region is required to be covered at any time. This is the first discussion of how to combine consideration of coverage and connectivity maintenance in a single activity scheduling. Tian et al. [38], enhance Zhang's work to support general wireless sensor networks by proving another conclusion: "the communication range is twice that of the sensing range" is the sufficient condition, and the tight lower bound to ensure that complete coverage preservation implies connectivity among active nodes if the original network topology (consisting of all the deployed nodes) is connected.

Energy conservation is a precious resource for battery-driven nodes in the network. Management of energy resources has significant impact on the ad-hoc network. Therefore, by handling the early drain of the battery node, controlling the transmission power of a node and putting low power consumption strategies together into the protocols, this management can prolong network lifetime [39]. Given a network that is connected when all nodes are operating at maximum power, the aim of power control is to use the minimum power level at each node for which the network remains connected [40]. A lot of achievements have been made in this subject. Hu [41] describes a fully distributed algorithm for topology control that provides reliability and graceful degradation under topological changes and component failures, which is based on a Delauney triangulation of the graph. Ramanathan and Rosales-Hain [42] present two distributed algorithms for mobile networks, named Local Information No Topology (LINT) and Local Information Link State Topology (LILT), to minimize the maximum transmission power. Rodoplu and Meng [43] describe a distributed position-based network protocol, and their algorithm is improved by Li and Halpern [44].

4.6.3 Dynamic lunar exploration robots routing protocol

To solve communication problems for the moon exploration, such as lack of sufficient hosts, limited power and high real-time requirement, we adopted multi-hop ad hoc wireless networks for our novel planetary exploration robots system (see Figure 4-11). Ad hoc wireless mobile networks are self-organizing and self-reconfiguring networks that can be established anytime and anywhere without the presence of static radio base stations or fixed backbone infrastructures. Then we developed an optimal ad hoc routing protocol [20, 45] named DLER (Dynamic Lunar Exploration Robots Routing Protocol) for our novel robot system. The project is supported by the National High Technology Research and Development Program of China (863 Program). The remains of this subsection will be organized as follow: first, we develop a node's energy consumption model based on analysis of the network topology of the communication system. Then the mathematical model for routing path optimization is built. Lastly, a new power-aware routing scheme is presented based on Dynamic Source Routing (DSR).

4.6.4 Energy consumption model

Since the most mobile hosts of an ad hoc network today operate using batteries, it is important to minimize the power consumption of the entire network. The power required by each mobile can be classified into

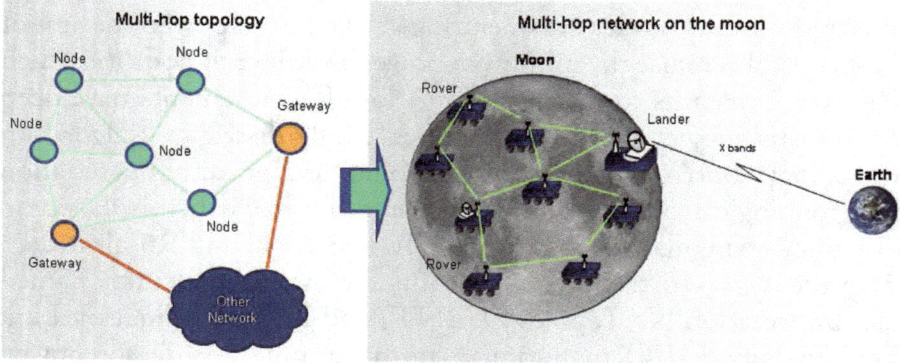

Figure 4-11 Conceptual frameworks of NOROS communication system. Source: [20] (with permission).

two categories: communication-related power and non-communication power. The former can be further divided into two parts, namely processing power and transceiver power [46]. According to different types of networks for different applications, the ratio of the power used for these two parts can be defined by:

$$\lambda = \frac{E_{NCR}}{E_{CR}} \tag{1}$$

where the denominator refers to the maximum power for the communication-related part and the numerator refers to the non-communication-related part.

There are four working modes of the network interface card, which are the doze, idle, receiving and transmitting modes. The power of each working mode is represented as P_{doze}, P_{idle}, P_{rcvd}, P_{tx}, and T_{doze}, T_{idle}, T_{rcvd}, T_{tx} is the time of each working mode taken at a node's service time T. Without consideration of the energy consumed during the working mode transformation, the total energy consumption of a NIC can be formulated as follows:

$$E_{NIC} = P_{doze} \cdot T_{doze} + P_{idle} \cdot T_{idle} + P_{rcvd} \cdot T_{rcvd} + P_{tx} \cdot T_{tx} \tag{2}$$

This can be further extended as:

$$E_{NIC} = P_{doze} \cdot T_{doze} + P_{idle} \cdot T_{idle} + P_{rcvd} \cdot \sum_{i=1}^{N} \frac{S_{rcvd}(i)}{D_{rcvd}} + P_{tx} \cdot \sum_{i=1}^{M} \frac{S_{tx}(i)}{D_{tx}}$$

where S(i) is the size of packet i, N and M are the number of packets received and transmitted, and D_{rcvd} and D_{tx} are the receiving frequency and transmitting frequency, respectively.

P_{rcvd} is the power used to transmit the packets from the source to the destination node, which can be extended as:

$$P_{rcvd} = P_{tx} \cdot B_{tx} \cdot G_{tx} \cdot L_{path} \cdot G_{rcvd}$$

where B_{tx} is the bandwidth of the NIC, and G_{tx} and G_{rcvd} are the gain of the sending antenna and receiving antenna separately. L_{path} is the propagation loss, which is calculated in the free space model.

As the energy consumption of each mobile related to the communication is mainly caused by the NIC working, consequently the energy consumed by each mobile node can be modeled as:

$$E_{node} = E_{CR} + E_{NCR} = (1 + \lambda)E_{CR} \approx (1 + \lambda)E_{NIC}$$

Generally we assume that P_{doze} and P_{idle} are negligibly small. Finally we get the energy consumption model of the mobile node as:

$$E_{node} = (1+\lambda)\left[P_{rcvd} \cdot \sum_{i=1}^{N} \frac{S_{rcvd}(i)}{D_{rcvd}} + P_{tx} \cdot \sum_{i=1}^{M} \frac{S_{tx}(i)}{D_{tx}}\right] \quad (3)$$

4.6.5 The optimal ad hoc routing protocols design for multi moon exploration robots system

The environmental conditions on the Moon are extremely bad. Missions like long-range surface exploration or an extended stay are challenging. This requires the communication system to be more robust, flexible and energy saving. Concretely, the characteristics can be summarized as:

- Lack of sufficient host station supporting
- Limited number of surface rovers
- Large communication packet size and high transition frequency
- Real-time communication request for commands like sudden stop
- Limited power sources with long lifetime communication networks required.

To satisfy the special requirements mentioned above, we designed an optimal routing algorithm, which selects the best path to minimize the total power needed to route packets on the network, and to maximize the lifetime of all nodes. Considering the integrity of the multi robots system, the lifetime of the networks is defined as the duration until the first occurrence of any node's power down. The optimization problem of selecting the best path can be described as:

$$\begin{aligned} &\min \sum_{i=1}^{d-1} E(n_i, n_{i+1}) \\ &\text{s.t.}\quad d < c \\ &\qquad E_{rest} > E_{min} \end{aligned} \tag{4}$$

where $E(n_i, n_{i+1})$ is the transmission power between hosts n_i and n_{i+1}, d is the routing hop count, E_{rest} is the minimum battery power reserved during the whole nodes, and c is a constant which is negatively correlated with the real-time request. To avoid unexpected power down, there should be a certain level of E_{min}.

The above function can be solved by the standard shortest path algorithm such as the Dijkstra or Bellman-Ford algorithm. Dijkstra's shortest path algorithm was modified to obtain the minimum total power route (MTPR) [47]. However, since the transmission power depends on the distance, this algorithm needs select routes with more hops than other routing-algorithms. With more nodes present in a route, the greater the end-to-end delay. Also, such a route is more likely to be unstable, because the probability that intermediate nodes will migrate will be higher. Hence, from the standpoint of minimum hops, the route obtained from the above algorithm is not attractive. We utilize the idea of DSR (Dynamic Source Routing), where the head of the packet contains the information of all routed nodes. The constraints are balanced between energy consumption and energy which must be reserved. When all nodes are surplus of energy, the count of hops is the main index to optimize the routing method, while some nodes' energy is deficient. To ensure that these nodes will not be overused to power down, the reserved energy constraint should be considered. The total transmission power for route P can be defined as follows, and details can be found in [45]:

$$X = \sum_{i=1}^{d-1} P(n_i, n_{i+1})$$

And the strategy for optimized route can be described as:

$$\begin{aligned}
&\text{define } X = \sum_{i=1}^{d-1} P(n_i, n_{i+1}) \\
&\text{define } \theta = \left\{ P \mid P \in \Omega \text{ and } \min_{n_i \in P} E_{n_i} \geq E_{min} \right\} \\
&\text{if } \theta \neq \emptyset, \text{ then } R_{energy_{aware}} = \arg\min_{P \in \theta} \{X + \alpha d\} \\
&\text{if } \theta \neq \emptyset, \text{ then } R_{energy_{aware}} = \arg\max_{P \in \Omega} \left\{ \min_{n_i \in P} \{E_{n_i}\} \right\}
\end{aligned} \tag{5}$$

where θ is the set of executed routes, Ω is the set containing all possible routes, d is the hop count of route P, E_{n_i} is the energy reserved for node n_i, and α is a constant is used to modify the requirement of real-time.

We adopted a modified distributed Bellman-Ford algorithm [48] to solve this optimization problem. At node n_i, it computes as:

$$C_{i,j} = Cost(n_i) + P_{transceiver}(n_i) + P_{transmit}(n_i, n_j)$$

where n_i is a neighbor node of n_j, $P_{transceiver}(n_i)$ is the transceiver power at node (n_i), and $Cost(n_i)$ is the total power cost from the source node to node n_i. This value is sent to node n_j. Subsequently, at node n_i it computes its power cost using:

$$\begin{aligned}
&Cost(n_j) = \min_{n_i \in \varphi}(C_{i,j} + \alpha j) \\
&\text{where } \varphi = \{n_i \mid n_i \text{ is a neighbor node of } n_j\}
\end{aligned}$$

The path with minimum cost from the source node to node n_j is chosen. This procedure is further repeated until the destination node is reached. Therefore, in this algorithm, $P_{transceiver}(n_i)$ helps the algorithm to find routes with fewer hops [49].

4.6.6 Simulation and results

The simulation of our routing protocol is done using OPNET software. To evaluate the impact of different route selection schemes on the lifetime of mobile hosts and the real-time performance, three cases are considered in our simulations, based on DSR protocols, DLER protocols

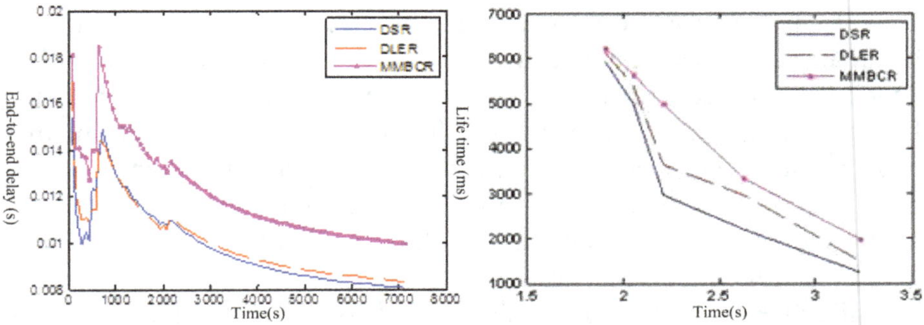

Figure 4-12 The results of various routing protocol simulation. Source: [45] (with permission).

and MMBCR (Min-Max Battery Cost Routing) protocols separately. We record the end-to-end delay, which is seen as the evaluation criterion of the real-time performance, and the expiration time of each node. The results of the simulation are shown below (Figure 4-12).

From the above, it can be determined that the end-to-end delay tends to be a small and stable value. Consequently, the real-time performance of three protocols are all relatively good, and the DSR protocols perform better because the smallest hop counts. Considering the lifetime of the networks, the MMBCR performs best at the cost of real-time performance. In order to ensure real-time performance, the DSR scheme always chooses the route which contains the smallest hop counts, although it may cause the battery power of some nodes to be consumed rapidly. The DLER scheme is a compromise, which takes the lifetime of the networks and the real-time performance into consideration simultaneously. It is more suitable for the application of a multi-robots system for moon exploration.

4.7 Perception challenges and achievements

4.7.1 Key problems

A variety of mechanical, thermal, biological, chemical, optical and magnetic sensors can be attached to mobile robots to measure properties of the environment. Unlike a fixed sensor network, which can only collect data at fixed positions in space, the mobile sensor networks have the ability to reposition and organize themselves in the network. For example, when an event is detected at a specific location, it is important to direct more mobile sensors toward the location to observe the event for more information. A natural question is: which mobiles should be sent out, and where should the nodes be placed in order to maximize the quality of the estimates? Since there is a cost associated with mobile movement and data transmission, the cost and the quality of perception should be balanced. Another key difference is data distribution. In a static wireless sensor network, data can be distributed using fixed routing or flooding, while dynamic routing is used in a mobile wireless sensor network, as mentioned in Chapter 2 of the companion book Mobile Robots for Dynamic Environments.

While there are many new probabilities brought on by the mobility of the platform, a lot of challenges are still to be resolved. For example, consider the problem of estimating the state of the network. State estimation requires the estimation of the state of robots and the environment based on local, limited-range sensory information. Localization of n vehicles in an m-dimensional configuration space requires $O((mn)^k)$ computations, where k is somewhere between 3 and 6, depending on the algorithm and domain-specific assumptions. The estimation problem is further exacerbated by the fact that not all robots in the network may be able to get the necessary information in a time-critical fashion. For better use of network communication and computation resources, in-network data processing techniques have been proposed. These enable the network to compute accurate and up-to-date global pictures of the global perception landscape that are available to all the robots in the system. An artificial potential field method was used for computing distributed maps in perception space, in order to compute adaptive paths for a mobile node that can interact with the sensor network [19]. Other methods such as gradient computations, particle filters and Bayesian inference are also included.

4.7.2 Major accomplishments

One of the fundamental issues in sensor networks is the coverage problem, which has been defined as the maximization of the total area covered by the robot's sensors [50]. In general, this reflects how well an area is monitored or tracked by sensors. This problem can be formulated as the Art Gallery Problem, which is to determine the number of observers necessary to cover an art gallery such that every point in the art gallery is monitored by at least one observer [51]. This problem can be solved optimally in a 2D plane, but is shown to be NP-hard when extended to a 3D space [52]. Meguerdichian et al. [53], proposed an optimal polynomial time algorithm that uses graph theoretic and computational geometry constructs for solving for best and worst case coverage. Some research attempted coverage of an initially unknown environment for mobile robots [54]. However, when the geometry of the environment is known in advance, coverage becomes a special case of path planning [55]. How to find the minimal and maximal exposure path that takes into account the duration that an object is monitored by sensors has been addressed [56]. A variant which allows the use of mobile sensors is known as the watchmen tours problem. In these approaches the sensor model is abstract and not well-suited to real environments and cameras. Some control strategies are based on complete knowledge of the event distribution density, and the control law positions the mobile sensors optimally with respect to the event distribution density function. These approaches guarantee the network minimizes a cost function relevant to the coverage problem locally, yet it is not reactive to the sensed environment. Other strategies [57, 58] are proposed based on node estimates, and they require that each agent can measure the value and gradient of the distribution density function at its own position, to maintain or seek a near-optimal sensing configuration. These strategies provide a significant reduction in computational overhead for each mobile, by eliminating the need for the numerical integration of a function over a polynomial domain at every time step.

Another fundamental issue in sensor networks is the localization problem. The overwhelming reason is that a sensor's location must be known for its data to be meaningful. As an additional motivation, sensor location information (if accurate enough) can be extremely useful for scalable "geographic" routing algorithms [59]. Location information is needed to track the placement of the nodes and to correlate the values measured by the nodes with their physical location. However, putting GPS receivers in every node or manually configuring locations is not cost-effective for most mobile sensor network applications. A lot of algorithms have been proposed for the localization of the wireless sensor networks.

We consider a network composed of anchor nodes and blind nodes. According to the mobility of the nodes, the network can be treated in three kinds of scenarios as follows,

1. Blind nodes are static while anchor nodes are moving. For example, a military application where nodes are dropped from a plane onto land, and transmitters attached to soldiers or animals in the area are used as moving seeds. Each node's location

estimate should become more accurate as time passes and it receives information from more seeds;

2. Blind nodes are moving while anchor nodes are static. One example is NASA's Mars Tumbleweed project. It proposes a low cost way to explore large areas on Mars by having rovers with sensors that are blown over the surface by the wind, with minimal or no control over their movement. Of course, GPS does not work on Mars, but it may be possible to establish fixed landmark seeds or positioning from orbiters [60].
3. Both blind nodes and anchor nodes are moving. This is the most general situation. It applies to any application where the nodes are deployed in an ad hoc way, and move either because of the environment they are in (wind, currents, etc.) or because they have actuators for motion.

According to the characteristics of the mobile wireless sensor network, the methods for localization can be classified into two broad classes: centralized localization techniques depend on sensor nodes transmitting data to a central location, where computation is performed to determine the location of each node; distributed localization methods do not require centralized computation, and rely on each node determining its location with only limited communication with nearby nodes. The distributed localization methods can be further classified into range-based and range-free methods. Range-based techniques use distance estimates or angle estimates in location calculations, while a range-free solution depends only on the contents of received messages. The main range-based approaches are illustrated as follows:

1. Angle of Arrival Detection Method (AOA) [61] – This method uses the antenna to directly detect the range and bearing of a target. To detect a target's range and bearing, a complicated and expensive antenna is installed in a fixed location. This antenna rotates, so when the radio waves coming from the target are at peak power, the antenna is able to determine the target's range and bearing. The advantage of using this method is the high level of accuracy it achieves.
2. Time/Time Difference of Arrival Detection Method (TOA/TDOA) [62, 63] – In this method, the sender transmits radio signals. After the radio signal reaches the target, the time spent

in transit is converted into the distance traveled to find the distance between the sender and target. From this, the target's range and bearing can be determined. However, if this method is used indoors, obstacles may interfere with the radio waves, lowering the method's accuracy.

3. Received Signal Strength Method (RSS) [64] – In this method, the sender transmits a radio signal. When the signal reaches the antenna at the target, the power of the received signal can be converted into distance to give the distance from the sender to the target as well as the target's range and bearing.

The main range-free approaches are as introduced as follows:

1. Centroid Scheme [65] – Centroid scheme is a range-free, proximity-based, coarse-grained localization algorithm that uses anchor beacons, containing location information, to estimate node position. In this method the receiver localizes itself with high confidence (under an idealized radio model) to the centroid of a set of proximate reference points using a connectivity metric. Each node estimates its location by calculating the center of the locations of all anchors it hears. If anchors are well positioned, location error can be reduced.
2. APIT Scheme [66] – APIT (Approximate Point In-Triangulation Test) employs a novel area-based approach to perform location estimation by isolating the environment into triangular regions between beaconing nodes. A node's presence inside or outside of these triangular regions allows a node to narrow down the area in which it can potentially reside. By utilizing combinations of anchor positions, the diameter of the estimated area in which a node resides can be reduced, to provide a good location estimate.
3. DV-Hop Scheme [61] – DV-Hop localization uses a mechanism that is similar to classical distance vector routing. In this method, one anchor broadcasts a beacon to be flooded throughout the network containing the anchors' location with a hop-count parameter initialized to one. Each receiving node maintains the minimum counter value per anchor of all beacons it receives and ignores those beacons with higher hop-count values. Beacons are flooded outward with hop-count values incremented at every

intermediate hop. Through this mechanism, all nodes in the network (including other anchors) get the shortest distance, in hops, to every anchor. This method provides localization in networks where seed density is low. Hop-counting techniques propagate location announcements throughout the network.

4. Amorphous Scheme [67] – This algorithm, proposed independently from DV-Hop, uses a similar algorithm for estimating position. First, like DV-Hop, each node obtains the hop distance to distributed anchors through beacon propagation. Once anchor estimates are collected, the hop distance estimation is obtained through local averaging. Each node collects neighboring nodes' hop distance estimates and computes an average of all its neighbors' values. Half of the radio range is then deducted from this average to compensate for error caused by low resolution. The Amorphous Localization algorithm takes a different approach from the DV-Hop algorithm to estimate the average distance of a single hop. The coordinates of anchors are flooded throughout the network so each node can maintain a hop-count to that anchor. Nodes calculate their position based on the received anchor locations and corresponding hop count.

None of these schemes target the case where anchors and blind nodes can both move. Though these methods can be adapted for mobile networks by refreshing location estimates frequently, they are not designed with any consideration for how mobility can be exploited to achieve localization. The methods that consider localization with mobile nodes are not very mature now. Bergamo and Mazzini's work [68] was the first approach that considered mobility of nodes. They assumed a network with two seeds at fixed locations that can transmit across the entire network, and nodes that are able to measure signal strength accurately. Instead of using mobility to improve localization, they studied how mobility makes localization more difficult, and found that errors increased with increasing node speed. Hu and Evans [69] first introduced the sequential Monte Carlo method into sensor networks localization, and argued that it can exploit mobility to improve the accuracy and precision of localization. This approach does not require additional hardware on the nodes and works even when the movement of seeds and nodes is uncontrollable. There are a lot of methods developed based on this work, such as

[70–72]. There are also some localization methods based on robotic localization, which usually assumes a prior or previously learned map, and tries to determine the robot's position based on its motion and sensor data [73, 74]. However, these methods can impose a significant memory and computational burden, especially if one is interested in high resolution.

4.7.3 Dynamic CSS localization system for NOROS robots

We developed a localization and navigation system, including software and hardware (see Figure 4-13), for our NOROS robots based on Chirp Spread Spectrum technology [75], and then in terms of M Cao's work [76], proposed an approach to correct distance measurement errors for CSS-based localization, through the combination of distance geometry constraints and a CSS ranging error model.

The major advantage of the CSS technology is to provide robust performance for LR-WPAN (low rate wireless personal areal network), even in the presence of path loss, multi-path, and reflection [77]. The key localization technology used in CSS is called symmetric double side two way ranging (SDS-TWR). SDSTWR does not require any kind of time synchronization between the transmitter and the receiver; instead it only depends on time-of-flight between two devices in order to estimate the range [78]. Cao introduces the Cayley–Menger determinant as an important tool for formulating the geometric relations among node positions in sensor networks as quadratic constraints, and estimates the errors in the inaccurate measured distances between sensor nodes and anchor nodes [76]. This method can be seen as optimizing the estimation of error by Least Square Method, and provides an optimal localization estimation, but doesn't perform well under line-of-sight environments. We developed Cao's method by modifying the parameters of error estimation equation

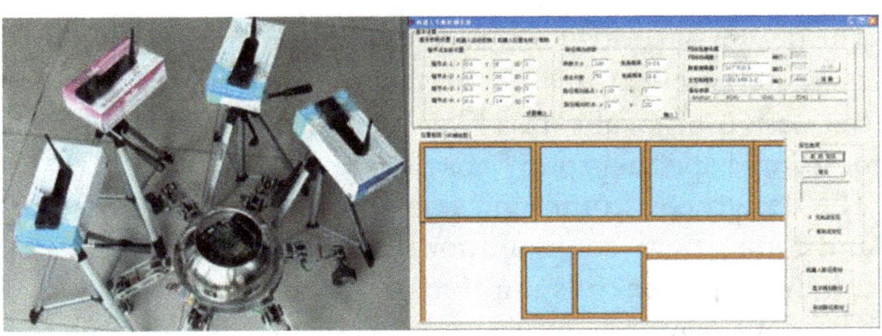

Figure 4-13. (a) Hardware of NOROS localization system (b) software of NOROS localization system.

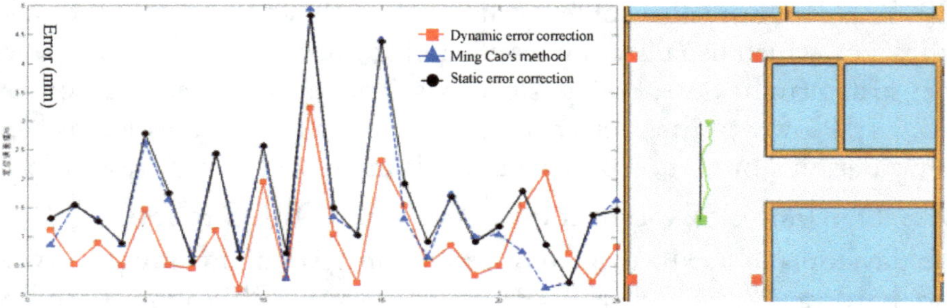

Figure 4-14 (a) The error estimation results of three algorithms (b) The tracking results of walking in line.

dynamically, and the experiments resulted in contrast with the other two mentioned methods. Figure 4-14(a) shows the dynamic error correction algorithm we proposed as having a good performance at dynamic error estimation. The results of the navigation and localization experiment are presented in Figure 4-14(b), where the black line is the actual path, and the green line is the path estimate by our localization system.

4.8 Conclusions and future works

This chapter presents the main problems and achievements of networked control mobile robots, which are embodied on the intersection between networked control, communication and perception. The developments of networked control mobile robots offer significant potential for such applications as environmental monitoring, surveillance, industry and security. However, there are still many challenges to be solved. The main overarching challenges are summarized here:

- The design of control methods to overcome drawbacks due to the characteristic features of the communication method adopted; the design of local control law to coordinate multiple robots to perform a special group behavior;
- The design of a communication protocol to guarantee coverage and connectivity configurations; the optimized routing design to minimize power consumption of the entire network.
- The coverage and localization problems using wireless sensor networks.

Many related works, such as the study of group behaviors, optimal routing design and wireless sensor network localization have been carried out using our NOROS robots system. The results encourage us greatly regarding networked control for mobile robots. Further study will focus on creating self-organizing robot networks with completely decentralized controllers and estimators, and finding new mathematical tools to solve the changing topology models of mobile robots. The tremendous potential of networked control mobile robots is still waiting for exploitation.

4.9 References

[1] Wiener, N., 1961, *Cybernetics: Or the Control and Communication in the Animal and the Machine*, The MIT press.

[2] Tsien, H. S., and Tsien, H., 1954, *Engineering Cybernetics*, McGraw-Hill, New York.

[3] Bemporad, A., Heemels, M., and Johansson, M., 2010, *Networked Control Systems*, Springer.

[4] Goldberg, K., 2001, *The Robot in the Garden: Telerobotics and Telepistemology in the Age of the Internet*, The MIT Press.

[5] Kunii, Y., Lee, J., Ohashi, T., Ando, N., Nakai, H., and Hashimoto, H., "Robotic network system: Mobile robot as a physical agent," *Proc. in Proc. 15th RSJ Workshop*.

[6] Kumar, V., Rus, D., and Sukhatme, G., 2008, "Networked Robots," *Springer Handbook of Robotics*, B. Siciliano, and O. Khatib, eds., Springer, Berlin Heidelberg, pp. 943–958.

[7] Hammond, G., and Hammond, G., 1986, "AGVS at work: Automated guided vehicle systems."

[8] Lee, J.-H., and Hashimoto, H., 2002, "Intelligent space—Concept and contents," *Advanced Robotics*, 16(3), pp. 265–280.

[9] Lee, J.-H., and Hashimoto, H., 2003, "Controlling mobile robots in distributed intelligent sensor network," *Industrial Electronics, IEEE Transactions on*, 50(5), pp. 890–902.

[10] Chaimowicz, L., Cowley, A., Gomez-Ibanez, D., Grocholsky, B., Hsieh, M., Hsu, H., Keller, J., Kumar, V., Swaminathan, R., and Taylor, C., 2005, "Deploying air-ground multi-robot teams

in urban environments," *Multi-Robot Systems From Swarms to Intelligent Automata Volume III*, Springer, pp. 223–234.

[11] Hsieh, M. A., Cowley, A., Keller, J. F., Chaimowicz, L., Grocholsky, B., Kumar, V., Taylor, C. J., Endo, Y., Arkin, R. C., and Jung, B., 2007, "Adaptive teams of autonomous aerial and ground robots for situational awareness," *Journal of Field Robotics*, 24(11–12), pp. 991–1014.

[12] Saucy, P., and Mondada, F., 2002, "KhepOnTheWeb: One year of access to a mobile robot on the Internet," *An Introduction to Online Robots*, The MIT Press, Cambridge, Massachusetts, London, England, pp. 99–115.

[13] McLurkin, J., and Smith, J., 2007, "Distributed algorithms for dispersion in indoor environments using a swarm of autonomous mobile robots," *Distributed Autonomous Robotic Systems 6*, Springer, pp. 399–408.

[14] Hagita, N., Kuwabara, K., and Tokuda, H., "Introduction to network robots," *Proc. of Workshop on Network Robot Systems, IEEE International Conference on Robotics and Automation (ICRA2005)*, pp. 1–4.

[15] Sukhatme, G. S., Dhariwal, A., Zhang, B., Oberg, C., Stauffer, B., and Caron, D. A., 2007, "Design and development of a wireless robotic networked aquatic microbial observing system," *Environmental Engineering Science*, 24(2), pp. 205–215.

[16] Kaiser, W. J., Pottie, G. J., Srivastava, M., Sukhatme, G. S., Villasenor, J., and Estrin, D., 2005, "Networked infomechanical systems (nims) for ambient intelligence," *Ambient Intelligence*, Springer, pp. 83–113.

[17] Rybski, P. E., Stoeter, S. A., Erickson, M. D., Gini, M., Hougen, D. F., and Papanikolopoulos, N., "A team of robotic agents for surveillance," *Proc. Proceedings of the Fourth International Conference on Autonomous Agents*, ACM, pp. 9–16.

[18] Kotay, K., Peterson, R., and Rus, D., "Experiments with robots and sensor networks for mapping and navigation," *Proc. Field and Service Robotics*, Springer, pp. 243–254.

[19] Li, Q., and Rus, D., 2005, "Navigation protocols in sensor networks," *ACM Transactions on Sensor Networks (TOSN)*, 1(1), pp. 3–35.

[20] Rovetta, A., and Ding, X., 2006, "Next steps for robotic landers rovers and outposts," *ILEWG 2006*, Beijing, China, pp. 23–27.

[21] Wang, Z., Ding, X., Xie, Y., Rovetta, A., Paul, C., and Espinosa, I., "Conceptual design of a novel robotics system for planetary exploration," Proc. Intelligent Control and Automation, 2006. WCICA 2006. The Sixth World Congress on, IEEE, pp. 8962–8965.

[22] Ding, X., and Xu, K., 2012, "Typical 3 + 3 Gait Motion Analysis of a Radial Symmetrical Six-Legged Robot Based on Metamorphic Theory," *Advances in Reconfigurable Mechanisms and Robots I*, Springer, pp. 535–550.

[23] Xu, K., and Ding, X., 2013, "Typical gait analysis of a six-legged robot in the context of metamorphic mechanism theory," Chinese *Journal of Mechanical Engineering*, 26(4), pp. 771–783.

[24] Franks, N. R., Pratt, S. C., Mallon, E. B., Britton, N. F., and Sumpter, D. J., 2002, "Information flow, opinion polling and collective intelligence in house–hunting social insects," *Philosophical Transactions of the Royal Society of London. Series B: Biological Sciences*, 357(1427), pp. 1567–1583.

[25] Hu, O., Tsui, P. W., Cragg, L., and Völker, N., 2004, "Agent architecture for multi-robot cooperation over the internet," *Integrated Computer-Aided Engineering*, 11(3), pp. 213–225.

[26] Tipsuwan, Y., and Chow, M.-Y., 2003, "Control methodologies in networked control systems," *Control Engineering Practice*, 11(10), pp. 1099–1111.

[27] Fattouh, A., and Sename, O., "H∞-based impedance control of teleoperation systems with time delay," *Proc. 4th Workshop on Time Delay Systems*, Citeseer.

[28] Witrant, E., De Wit, C. C., and Georges, D., "Remote output stabilization under two channels time-varying delays," *Proc. of the 4th IFAC Workshop on Time Delay Systems*.

[29] Seuret, A., Michaut, F., Richard, J.-P., and Divoux, T., "Networked control using GPS synchronization," *Proc. American Control Conference*, 2006, IEEE, p. 6.

[30] Vatanski, N., Georges, J.-P., Aubrun, C., Rondeau, E., and Jounela, S.-L. J., 2006, "Control compensation based on upper bound delay in networked control systems," arXiv preprint cs/0609151.

[31] Diouri, I., Georges, J.-P., and Rondeau, E., "Accommodation of delays for networked control systems using classification of service," *Proc. Networking, Sensing and Control, 2007 IEEE International Conference on*, IEEE, pp. 410–415.

[32] Castanon, D. A., and Wu, C., "Distributed algorithms for dynamic reassignment," *Proc. Decision and Control, 2003. Proceedings. 42nd IEEE Conference on*, IEEE, pp. 13–18.

[33] Moore, B. J., and Passino, K. M., 2007, "Distributed task assignment for mobile agents," *Automatic Control, IEEE Transactions on*, 52(4), pp. 749–753.

[34] Hsieh, M., and Kumar, V., "Pattern generation with multiple robots," *Proc. Robotics and Automation, 2006. ICRA 2006. Proceedings 2006 IEEE International Conference on*, IEEE, pp. 2442–2447.

[35] Ye, F., Zhong, G., Cheng, J., Lu, S., and Zhang, L., "PEAS: A robust energy conserving protocol for long-lived sensor networks," *Proc. Distributed Computing Systems, 2003. Proceedings. 23rd International Conference on*, IEEE, pp. 28–37.

[36] Wang, X., Xing, G., Zhang, Y., Lu, C., Pless, R., and Gill, C., "Integrated coverage and connectivity configuration in wireless sensor networks," *Proc. Proceedings of the 1st International Conference on Embedded Networked Sensor Systems*, ACM, pp. 28–39.

[37] Zhang, H., and Hou, J., "On deriving the upper bound of α-lifetime for large sensor networks," *Proc. Proceedings of the 5th ACM International Symposium on Mobile ad hoc Networking and Computing*, ACM, pp. 121–132.

[38] Tian, D., and Georganas, N. D., 2005, "Connectivity maintenance and coverage preservation in wireless sensor networks," *Ad Hoc Networks*, 3(6), pp. 744–761.

[39] Abu Bakar, K., Mohsin, A. H., Adekiigbe, A., and Ghafoor, K. Z., 2012, "A survey of energy-aware routing and MAC layer protocols in MANETS: Trends and challenges," *Network Protocols and Algorithms*, 4(2), pp. 82–107.

[40] Li, L., Halpern, J. Y., Bahl, P., Wang, Y.-M., and Wattenhofer, R., "Analysis of a cone-based distributed topology control algorithm for wireless multi-hop networks," *Proc. Proceedings of the Twentieth Annual ACM Symposium on Principles of Distributed Computing*, ACM, pp. 264–273.

[41] Hu, L., 1993, "Topology control for multihop packet radio networks," *Communications, IEEE Transactions on*, 41(10), pp. 1474–1481.

[42] Ramanathan, R., and Rosales-Hain, R., "Topology control of multihop wireless networks using transmit power adjustment," *Proc. INFOCOM 2000. Nineteenth Annual Joint Conference of the IEEE Computer and Communications Societies. Proceedings. IEEE*, *IEEE*, pp. 404–413.

[43] Rodoplu, V., and Meng, T. H., 1999, "Minimum energy mobile wireless networks," *Selected Areas in Communications, IEEE Journal on*, 17(8), pp. 1333–1344.

[44] Li, L., and Halpern, J. Y., "Minimum-energy mobile wireless networks revisited," *Proc. Communications, 2001. ICC 2001. IEEE International Conference on*, *IEEE*, pp. 278–283.

[45] Lu, Y., and Qin, S., 2009, "Optimization Design of Ad hoc Routing Protocol for the Communication of Lunar Exploration Multi? robots System," *Science & Technology Review*, 27(19), p. 19.

[46] Toh, C.-K., 2001, "Maximum battery life routing to support ubiquitous mobile computing in wireless ad hoc networks," *Communications Magazine, IEEE*, 39(6), pp. 138–147.

[47] Singh, S., and Raghavendra, C. S., 1998, "PAMAS—power aware multi-access protocol with signalling for ad hoc networks,"

ACM SIGCOMM Computer Communication Review, 28(3), pp. 5–26.

[48] Scott, K., and Bambos, N., "Routing and channel assignment for low power transmission in PCS," *Proc. Universal Personal Communications, 1996. Record., 1996 5th IEEE International Conference on, IEEE,* pp. 498–502.

[49] Toh, C.-K., Cobb, H., and Scott, D. A., "Performance evaluation of battery-life-aware routing schemes for wireless ad hoc networks," *Proc. Communications, 2001. ICC 2001. IEEE International Conference on, IEEE,* pp. 2824–2829.

[50] Gage, D. W., 1992, "Command control for many-robot systems," DTIC Document.

[51] O'rourke, J., 1987, *Art gallery theorems and algorithms,* Oxford University Press, Oxford.

[52] O'Rourke, J., 1992, "Computational geometry column 15," *International Journal of Computational Geometry & Applications,* 2(02), pp. 215–217.

[53] Meguerdichian, S., Koushanfar, F., Potkonjak, M., and Srivastava, M. B., "Coverage problems in wireless ad-hoc sensor networks," *Proc. INFOCOM 2001. Twentieth Annual Joint Conference of the IEEE Computer and Communications Societies. Proceedings. IEEE,* pp. 1380–1387 vol. 1383.

[54] Sukhatme, G. S., and Mataric, M. J., 2000, "Embedding robots into the Internet," *Communications of the ACM,* 43(5), pp. 67–73.

[55] Hofner, C., and Schmidt, G., 1995, "Path planning and guidance techniques for an autonomous mobile cleaning robot," *Robotics and Autonomous Systems,* 14(2), pp. 199–212.

[56] Veltri, G., Huang, Q., Qu, G., and Potkonjak, M., "Minimal and maximal exposure path algorithms for wireless embedded sensor networks," *Proc. Proceedings of the 1st international conference on Embedded Networked Sensor Systems,* ACM, pp. 40–50.

[57] Schwager, M., McLurkin, J., and Rus, D., "Distributed Coverage Control with Sensory Feedback for Networked Robots," *Proc. Robotics: Science and Systems.*

[58] Schwager, M., Rus, D., and Slotine, J.-J., 2009, "Decentralized, adaptive coverage control for networked robots," *The International Journal of Robotics Research,* 28(3), pp. 357–375.

[59] Patwari, N., Ash, J. N., Kyperountas, S., Hero III, A. O., Moses, R. L., and Correal, N. S., 2005, "Locating the nodes: Cooperative localization in wireless sensor networks," *Signal Processing Magazine, IEEE,* 22(4), pp. 54–69.

[60] Antol, J., Calhoun, P., Flick, J., Hajos, G. A., Kolacinski, R., Minton, D., Owens, R., and Parker, J., 2003, *Low cost mars surface exploration: The mars tumbleweed, Citeseer.*

[61] Niculescu, D., and Nath, B., "Ad hoc positioning system (APS)," *Proc. Global Telecommunications Conference, 2001. GLOBECOM'01. IEEE, IEEE,* pp. 2926–2931.

[62] Hofmann-Wellenhof, B., Lichtenegger, H., and Collins, J., 1993, "Global Positioning System. Theory and Practice," *Global Positioning System. Theory and practice,* by Hofmann-Wellenhof, B.; Lichtenegger, H.; Collins, J. Springer, Wien (Austria), 1993, 347 p., ISBN 3-211-82477-4, Price DM 79.00. ISBN 0-387-82477-4 (USA). 1.

[63] Savvides, A., Han, C.-C., and Strivastava, M. B., "Dynamic fine-grained localization in ad-hoc networks of sensors," *Proc. Proceedings of the 7th Annual International Conference on Mobile Computing and Networking,* ACM, pp. 166–179.

[64] Patwari, N., and Hero III, A. O., "Using proximity and quantized RSS for sensor localization in wireless networks," *Proc. Proceedings of the 2nd ACM International Conference on Wireless Sensor Networks and Applications,* ACM, pp. 20–29.

[65] Bulusu, N., Heidemann, J., and Estrin, D., 2000, "GPS-less low-cost outdoor localization for very small devices," *Personal Communications, IEEE,* 7(5), pp. 28–34.

[66] He, T., Huang, C., Blum, B. M., Stankovic, J. A., and Abdelzaher, T., "Range-free localization schemes for large scale sensor networks," Proc. *Proceedings of the 9th Annual International Conference on Mobile Computing and Networking, ACM*, pp. 81–95.

[67] Nagpal, R., 1999, "Organizing a global coordinate system from local information on an amorphous computer."

[68] Bergamo, P., and Mazzini, G., "Localization in sensor networks with fading and mobility," *Proc. Personal, Indoor and Mobile Radio Communications, 2002. The 13th IEEE International Symposium on, IEEE*, pp. 750–754.

[69] Hu, L., and Evans, D., "Localization for mobile sensor networks," *Proc. Proceedings of the 10th Annual International Conference on Mobile Computing and Networking*, ACM, pp. 45–57.

[70] Stevens-Navarro, E., Vivekanandan, V., and Wong, V. W., "Dual and mixture Monte Carlo localization algorithms for mobile wireless sensor networks," *Proc. Wireless Communications and Networking Conference, 2007. WCNC 2007. IEEE, IEEE*, pp. 4024–4028.

[71] Yi, J., Yang, S., and Cha, H., "Multi-hop-based monte carlo localization for mobile sensor networks," *Proc. Sensor, Mesh and Ad Hoc Communications and Networks, 2007. SECON'07. 4th Annual IEEE Communications Society Conference on, IEEE*, pp. 162–171.

[72] Wang, W., and Zhu, Q., "Varying the sample number for Monte Carlo localization in mobile sensor networks," *Proc. Computer and Computational Sciences, 2007. IMSCCS 2007. Second International Multi-Symposiums on, IEEE*, pp. 490–495.

[73] Burgard, W., Fox, D., Hennig, D., and Schmidt, T., "Estimating the absolute position of a mobile robot using position probability grids," *Proc. Proceedings of the National Conference on Artificial intelligence*, pp. 896–901.

[74] Burgard, W., Derr, A., Fox, D., and Cremers, A. B., "Integrating global position estimation and position tracking for mobile robots: The Dynamic Markov Localization approach," *Proc.*

Intelligent Robots and Systems, 1998. Proceedings, 1998 IEEE/RSJ International Conference on, IEEE, pp. 730–735.

[75] Xi, T., and Ding, X., 2013, "Dynamic Error Correction Algorithm Based on CSS Localization," *Jiangsu Machine Building & Automation*, 42(1), pp. 89–92.

[76] Cao, M., Anderson, B., and Morse, A. S., 2006, "Sensor network localization with imprecise distances," *Systems & Control Letters*, 55(11), pp. 887–893.

[77] Ahn, H.-S., Hur, H., and Choi, W.-S., "One-way ranging technique for CSS-based indoor localization," *Proc. Industrial Informatics, 2008. INDIN 2008. 6th IEEE International Conference on, IEEE*, pp. 1513–1518.

[78] Chan, Y.-T., Tsui, W.-Y., So, H.-C., and Ching, P.-C., 2006, "Time-of-arrival based localization under NLOS conditions," *Vehicular Technology, IEEE Transactions on*, 55(1), pp. 17–24.

5. Human-machine interface of mobile robot for posture

I-Ming Chen, Wei-Ting Yang, Albert Causo, and Song Huat Yeo, Nanyang Technological University, Singapore

Abstract: This chapter presents a human-machine interface system for mobile robots to capture and replicate human posture. A motion replication system (MoRep) that captures whole body motion using inertia sensor-based measurement units is presented. Multimodal feedback enables users to replicate measured postures.

5.1 A survey of HMI for robots

A human-machine interface of a mobile robot is supposed to facilitate the operation of mobile robots in performing tasks in remote environments. Mobile robots have been around for quite some time now, and various control interfaces have been designed and implemented for these machines [1]. Many factors influence the design of an interface, such as structure of the mobile robot, complexity of task, accuracy of task, and situation of operator [2–4]. A good human-machine interface allows an operator to accomplish tasks quickly, smoothly and accurately in manual remote or tele-operation status. Technologies such as balance control, motion control, path planning and obstacle avoiding aid in decreasing the complexity of operations to a certain extent [5–8]. The operator can now pay attention to key functions and operations while performing tasks. The design of human-machine interfaces could also be simplified and become more intuitive.

A human-machine interface (HMI) or human-computer interface (HCI) usually includes software and hardware components [3, 9]. The software component includes an operating window on a screen, which could show virtual buttons, readings, videos, images and graphs. The software part is designed to aid in operation, monitoring and reminding via the operator's vision. The hardware component includes devices to receive commands or signals from an operator via physical contact. The hardware part of a mobile robot is usually composed of more than one device. Numerous human-machine interfaces have been developed to cater to different mobile robot platforms and requirements [2, 10–12]. They could be roughly classified into five types (Figure 5-1):

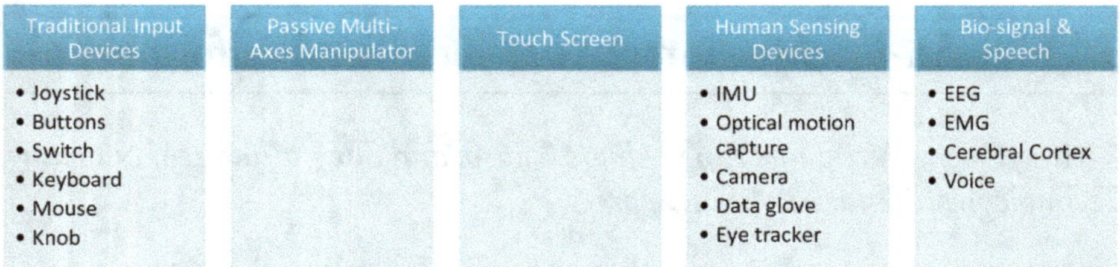

Figure 5-1 Five types of human-machine interfaces.

5.1.1 Traditional input devices

Most traditional input devices are standard components such as buttons, knobs, switches, keyboards, mice and joysticks. The early human-machine interface was a control panel with these input devices, and had the advantages of high stability, high strength, and low cost. This kind of interface is suitable for mobile control in hazardous operating environments, but the tasks need high stability. These days, these traditional interfaces have been adapted to industrial, military and security robots [13]. However, these interfaces provide insufficient control for mobile robots.

5.1.2 Passive multi-axes manipulator

The robot arm is an important part attached to a tracked, legged or wheeled mobile platform in order to perform complex tasks [1]. Some robot arms adopted a passive manipulator as human-machine interface (Figure 5-2), similar to an advanced joystick. The structure of this device is like a small robot arm and its figure is similar, with the robot arm we want to control. Passive manipulator devices make operation of the robot arm more intuitive. Sensors are installed on the joints of the

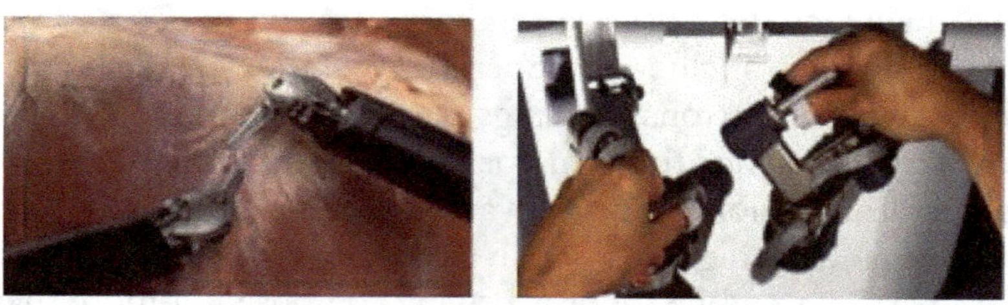

Figure 5-2 The robot arm (left) and operation manipulation (right) of Da Vinci surgical system. Source: [14] with permission, ©2014 Intuitive Surgical, Inc.

manipulator to measure displacement, velocity, rotational angle or rotational velocity. Thus, a robot arm could be controlled and driven by the processed signals of sensors. The design of this interface is quite suitable for a robot arm to perform tasks that need accurate manual teleoperation, as with surgical robotics and bomb-disposal robots. Moreover, passive manipulators could also convey resistance and vibration to an operator as the robot arm encounters resistance or obstacles [15, 16]. This is a mechanism called haptic feedback [17]. The mechanism is realized by installing servomotors or adjustable dampers in the joints of a passive manipulator. The magnitude of resistance is determined by the values of the force sensors attached to the robot arm. Haptic feedback helps an operator to properly control the output force and avoid collision. PHANTOM haptic devices (SensAble Technologies, Inc.) are innovative platforms of a passive multi-axes manipulator with haptic feedback, and is used in various applications [18, 19].

5.1.3 Touch screen

The advent of mobile devices such as phones and tablets made touch screen a popular choice for interface devices, owing to its portability and high performance. It may gradually replace some traditional input devices such as buttons and knobs. However, touch screen is not appropriate when it comes to controlling complex machinery and motion. For guide robots, found in buildings, touch screen is very important, as users can directly input data into the robots without the use of external input devices such as keyboards or mice.

5.1.4 Human motion sensing devices

Any sensor systems that could detect or capture human motion are included in this category. For example, optical motion capture systems, inertial measurement units (IMUs), smart gloves, cameras and eye trackers are all human motion sensing devices. In recent years, novel human motion sensing devices have been developed and applied to many fields [20–24]. When these devices are applied to the human-machine interface of a mobile robot, the operator himself becomes the remote controller. The interface cannot only realize simple control by recognizing specific motion and gesture, but also control complex humanoid robots by capturing human motion [21, 24–27]. Human motion sensing interface is nimble, flexible, and has wide applications.

5.1.5 Bio-signal capture and speech recognition system

Electrodes could capture bio-signals such as EEG and ECG, and many researchers attempt to convert these multi-channel signals into stable and definite control signals [28–30]. This robotics interface is usually employed to aid special operators such as the elderly, people with disabilities or paralysis patients [31]. This interface could be used in assistive technologies like prostheses, wheelchairs and exoskeleton devices. In order to obtain stable bio-signals for accurate and complex control, some researches even capture signals from the cerebral cortex and muscle nerves by surgical operation [32]. Speech recognition is a technology that has been around for quite some time [33]. Since sound is a single channel signal, for a computer to recognize and understand human language it needs a sophisticated algorithm and a single microphone. Speech recognition had been applied to robot control several years ago [34]. However, speech recognition is still not a robust and safe input method for mobile robots. It is not easy to process speech content due to speaker accent and noisy environment.

5.1.6 Challenges and open problems of HMI

5.1.6.1 Challenges of current HMI

Although the demand for HMI has multiplied over the years, some basic challenges still remain. Table 5-1 lists some of the drawbacks for each of the HMI categories listed above. Traditional input devices tend to be passive and unnatural in certain conditions, like in the use of keyboards for navigation control. While passive multi-axis manipulators allow for better control of robots, the device is suitable mostly for arm-like robots only.

Although touch screen has found a niche in controlling industrial robots and concierge robots, it cannot be used for applications that require wide-area interfaces due to cost. The rise of human motion sensing devices can be attributed to the lower cost of sensors. However, data from motion sensing devices require considerable signal processing and calibration to be reliable. Moreover, only the most expensive sensors tend to have high sensitivity and resolution. Bio-signals and speech recognition, offer the most promise in terms of enabling natural HMI. However, it is also the most noise-sensitive and unstable of all the HMIs. Data from bio-signal and speech sensors require noise cleaning and extensive processing to be useful.

Table 5-1 Summary of HMI challenges.

HMI	Challenges
Traditional Input Devices	Provides inadequate way to control the robot Unnatural interface Mostly passive devices
Passive Multi-Axes Manipulator	Bulky devices Expensive Appropriate mostly to arm-like robots
Touch Screen	Limited size due to price Cheap screens may not be responsive enough Passive device
Human Motion Sensing Devices	Sensing stability could be an issue May require considerable signal processing before data becomes usable Sensitivity and resolution may not be adequate Sensor packaging may make it difficult for users to wear
Bio-Signal Capture and Speech Recognition System	Requires extensive signal processing before data becomes usable Sensors are easily affected by noise Signal stability could be an issue

5.1.6.2 Open problems of HMI

The challenges listed above are just some of the factors that need to be considered when designing HMI for a mobile robot. In fact, recent advances have also exposed persistent issues that are yet to be addressed in a concerted effort by researchers. Some of these open problems include:

- **Managing cognitive overload**
  With the availability of multitude of sensors, users may get overwhelmed from the deluge of data, which they have to process while controlling a robot. A systematic study on this issue is

normally conducted within the social and medical sciences; however, findings in those fields are yet to fully benefit the robotics community.

- **Designing challenge**
  With many options on which HMI to implement, general guidelines are needed. Factors that need to be considered in choosing the HMI design should include the type of robot to control or manipulate, the environment at which the user will operate, and the speed, reliability, and sensitivity requirements needed. In HMI like human-motion sensing, issues like sensor positioning and number of sensors are also critical factors.

- **Environment sensing**
  So far, most of the focus of HMIs has involved just the robot and the user. The environment is either assumed or simplified in order to make the system work. However, the need for an HMI that fully recognizes the environment at which the user and the robot operate should be factored in as well.

The rest of this chapter will discuss one type of HMI that we have focused on – human-motion sensing based systems. The measurement of human upper arm motion, combined with the use of multi-modal feedback, allows users to have an HMI experience that enables motion learning and replication. Some of the challenges mentioned above, as well as the open problems in HMI, can actually be seen in operation of this system.

5.2 Motion replication System

Human-machine interface has to be re-thought. In the past, human-machine interface used for teaching and training robots was a one-way interaction. As technology and robotics advance, people are on the way to achieving two-way interaction, which means that the robot is also capable of teaching operators [35, 36]. People could input their knowledge to a robotic database, and the robot would become a platform to teach people. A good example is the humanoid telepresence robot [25]. A telepresence robot extends the presence of the human, which helps the human operator to attend class, meeting, or conference in a remote place. When using the robot, users must learn how to operate it at least once at the beginning. The robot could also teach the operator basic operations, as well as motions and gestures via the human-machine interface. Since the operator and the robot are in different places, the operator can learn motions and gestures from a virtual teacher, i.e., the robot itself. The concept of learning spatial body motion from a virtual teacher is called motion replication (MoRep). MoRep systems are getting to be an important part of human-machine interface, especially for humanoid robots [26, 27, 36]. Advanced humanoid robots usually need to perform accurate, complex and flexible tasks in hazardous environments, so the training of operators is crucial.

The basic structure of a MoRep system includes three main parts: motion capture, motion comparison and multi-modal feedbacks (see Figure 5-3). The system can capture user's motion and compare that to motions stored in a database. According to the result of the comparison, the system can send proper feedback to remind the user how to correct his/her movements. Human motion sensing devices are a critical component of motion capture.

Motion comparison is the heart of the MoRep system. There are various algorithms for the post-processing of the raw motion data, which

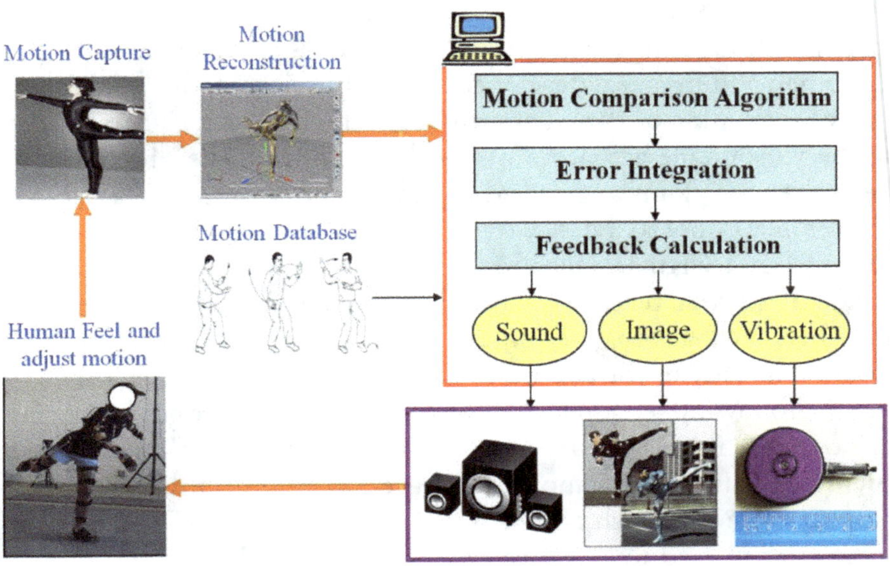

Figure 5-3 The structure of motion replication (MoRep) system.

may include positions, Euler angles, velocities or accelerations of human body segments; data generated are usually large. The first step is to extract useful information, which is used for similarity comparison. Tsuruta et al., proposed a virtual dance collaboration system to recognize body motion in real-time [37]. Fourteen feature values were extracted from the raw data using principal component analysis (PCA). Motion templates were constructed by training the feature values of several motions. The system recognizes the dancer's motion by comparing it with these templates. Hachimuea et al., apply the theory of Laban Movement Analysis (LMA) to extract characteristic poses and highlight parts from the data of dancing movement [38]. Baek et al., used curve fitting technique for motion comparison [39]. The curves were drawn by the chosen joint positions. By matching the turning points of the curves, the time-shifting problem was eliminated. The other points were moved by multilevel B-spline interpolation. The performance evaluation and feedback were based on the results of the curve fitting.

Feedback function is another critical component of the MoRep system. Vision, audio and tactile are three main types of communication used in learning motion. MoRep systems may adopt various combinations of feedback type [40, 41]. Baek et al., developed a virtual-reality based training system which contained an avatar motion-guiding interface, motion retargeting, and audio and image-advice feedback [39]. Lieberman and Breazeal developed a wearable vibrotactile feedback suit for motor training of the right arm [42]. The feedback signals for driving the vibrotactile actuators were calculated from the differences between the user's joint angles and the corresponding ones in a database. Chan et al., proposed a dance training system based on motion capture (MoCap) and virtual reality technology [43]. A student imitates the motion of a virtual teacher projected on the screen. The system provides immediate feedback to guide students to improve their skills by evaluating the similarity of the two motions.

In the following sections, a MoRep system for posture guidance of upper extremity is presented. An arm suit with two inertial measurement units (IMU) is used for motion capture. There is an intuitive human-computer interface with a 3D simulated environment. The orientation of the joint is used for posture comparison [44]. The system provides visual and verbal feedback to remind users how to correct their postures, which helps users to achieve required postures faster and to learn efficiently. Finally, the utility of this posture guiding system is tested in experiments.

5.3 Visual and verbal feedback

Two kinds of feedback are used in this MoRep system, visual and vibrotactile feedback. Choosing the right feedback makes learning the motion more intuitive and realistic.

5.3.1 Visual feedback

When a user wears the arm suit and starts the program, the left upper extremity is shown in 3D virtual reality. The interface window is split into two parts for displaying different views, as seen in Figure 5-4(a). The left window presents the lateral view, and the right window presents the anterior view. When the user starts to learn a posture, the virtual teacher's left upper extremity appears in the 3D virtual reality as a fixed posture goal. The user could achieve the assigned posture by driving the corresponding virtual extremity to superimpose the virtual teacher's. It is intuitive and fast to learn a new posture by using superimposition, which could be realized in the virtual world but not in the real world. A total score is shown on the top of the right window. The score indicates the similarity between teacher and user, and it ranges between 0 and 100. A larger score means a higher similarity. The total score is calculated by a comparison algorithm which will be discussed in the next section. The value of the total score is always updated to remind the user how close his/her posture is to the teacher's posture.

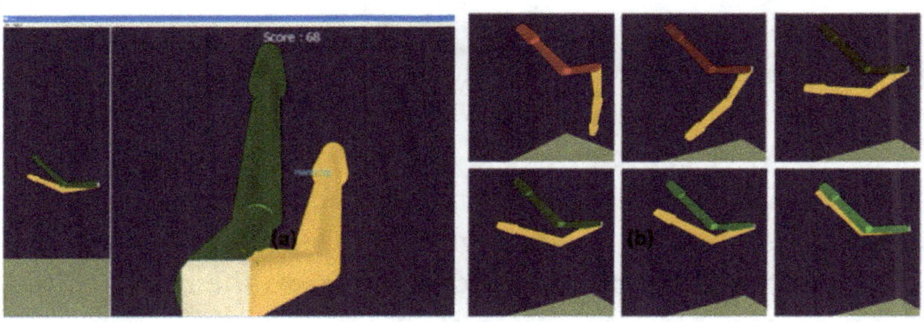

Figure 5-4 (a) Human-computer interface of the MoRep system. (b) The change of color of virtual teacher's upper extremity.

The color of the virtual teacher's extremity changes according to the similarity between the teacher's posture and the user's, as seen in Figure 5-4(b). The color is transparent dark red when the two postures are totally different, and it is transparent green when the two postures are the same. The rest of the states are presented by gradient colors between red and green. However, the color of the user's extremity is always bright yellow. This color arrangement avoids confusion. The change of color reminds the learner how close he is to the correct posture.

Besides the superimposition and the changes of total score and color, words reminding the learner how to move to achieve the assigned posture are shown in the middle of the right window. Words are updated when the motion direction of the user needs changing. This function helps the user to reach the required posture more systematically. When the words appear on screen, the system sends out the same verbal instructions. In fact, the words are determined through the verbal feedback strategy that is presented in the next section.

5.3.2 Verbal feedback

Verbal instruction is used for detailed adjustment, but it can become the main feedback when the learner cannot see the display due to occlusion. Good verbal instructions should be intuitive and correspond with human cognition and learning processes. The left upper extremity could be considered as a five degree of freedom (dof) model which includes the shoulder (3 dof), elbow (1 dof) and radius-ulna (1 dof) joint. The sequence of the adjustment and words for the feedback are defined according to this model. There are three steps necessary to achieve the goal posture: (1) adjust position of the wrist, (2) keep position of wrist and rotate whole upper extremity, and (3) rotate radius-ulna joint. Figure 5-5 shows the flowchart of feedback strategy. The arm suit captures the user posture, and the teacher's posture is stored in a database. After comparing the two postures, if the error is under default tolerance, no adjustment will be needed. If two postures are different, the user has to go through the three steps. Note that when the second or third step ends, the flow returns to the first step to check the previous state. If the condition is "No required adjustment" at the start of the third step, the flow goes back to compare the two postures again. This closed-loop checking strategy assures the accuracy of the user's posture.

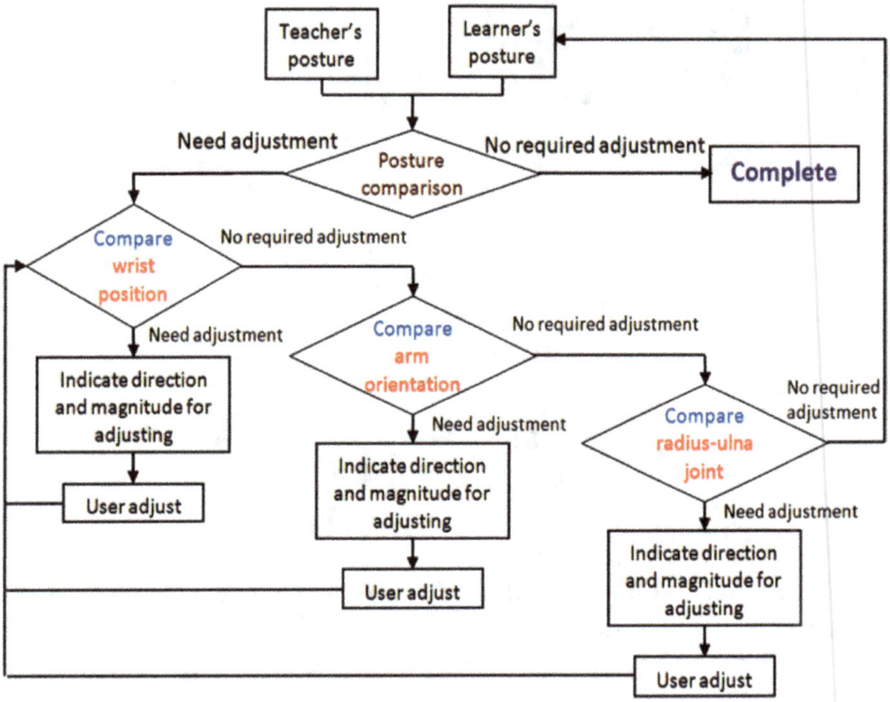

Figure 5-5 Flowchart of verbal feedback.

5.3.2.1 Verbal instruction for wrist position

According to human perception in motion [45], people usually learn new movements by remembering the position of the end or the joint, such as a hands or a knee. For the upper extremity, most of the required postures could be achieved once the wrist reaches the target position. Thus, the adjustment of wrist position is the first step. The upper left extremity of the user and the teacher are of the same size in the virtual reality, as seen in Figure 5-6(a). The 1×3 vector between the two wrists is called the error vector, V_d. It could be regarded as the direction to which the user should move to achieve the required posture. The length of the error vector is equal to the magnitude of movement. If the length exceeds one-twentieth length of the user's upper extremity, the verbal feedback will start.

The words to describe direction of spatial vector are usually ambiguous and not intuitive. Therefore, wrist motion is simplified into six basic

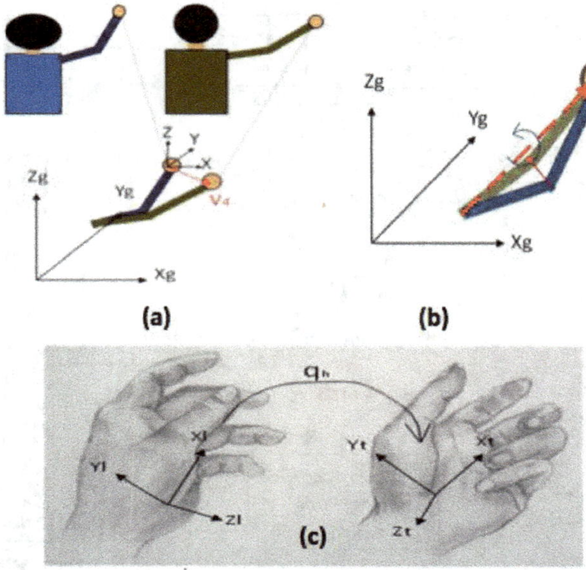

Figure 5-6 Diagram of two upper extremities in virtual world at (a) first step, (b) second step, and (c) third step.

directions: up, down, left, right, forward and backward. According to the values of elements of the error vector and their sign, proper direction is given as verbal feedback. Table 5-2 shows their mapping. At first, the absolute values of three elements are calculated, and then the maximum element is chosen. The direction of adjustment is chosen according to the original sign of the maximum element. The feedback direction is always updated by the state of the error vector. Once the length of the error vector is smaller than one-twentieth length of upper extremity, the procedure jumps to the second step.

In order to reduce the cognitive load on the user, the words used for verbal feedback are short, with clear meaning. The format is the word "hand" plus "direction of adjustment," such as "hand up" and "hand down." The directions right and left are replaced by inside and outside, since humans are not sensitive to the former.

5.3.2.2 Verbal instruction for arm orientation

The second step is to adjust the orientation of the whole upper extremity by fixing the wrist position. Figure 5-6(b) shows the diagram of the user's and teacher's upper extremities in virtual reality. After going

Table 5-2 Elements of error vector and corresponding directions.

$\vec{V}_d = (X_d, Y_d, Z_d)$	Sign	Directions
X_d	+	Forward
	–	Backward
Y_d	+	Left
	–	Right
Z_d	+	Up
	–	Down

through the first step, the positions of the two wrists and shoulders would coincide as of this step. The dash line is the vector of the rotation axis, which is from shoulder to wrist. The user could meet the required posture by rotating his upper extremity about the axis.

At this step, the similarity score of the arms is calculated first by comparison [44]. If the score is under 90, verbal feedback starts. Quaternion is used to represent the orientation of the segment. The required rotation axis and magnitude are calculated by using the arm orientation of user, q_{UArm}, and of teacher, q_{TArm}, as below:

$$\begin{aligned} q_r &= q_{UArm} \cdot (q_{TArm})^{-1} = (w, \vec{v}_r) \\ \theta_r &= 2\sin^{-1}(w) \end{aligned} \tag{1}$$

where v_r is the vector of the rotation axis and θ_r is rotation angle. Verbal instruction is determined by the direction of the rotation axis and the sign of the rotation angle. Positive v_r means its direction is from shoulder to wrist, and negative means the opposite. Table 5-3 shows the words

Table 5-3 Verbal instructions under different conditions at the second adjusting step.

Direction of v_r	Sign of Θ_r	Words of instruction
+	–	Keep hand in position,
–	+	rotate elbow inside
+	+	Keep hand in position,
–	–	rotate elbow outside

of instruction in all kinds of conditions. Rotation of the whole upper extremity is represented by "rotate elbow." The directions up and down are replaced by outside and inside respectively.

5.3.2.3 *Verbal instruction for hand orientation*

After going through the first and second adjusting step, there is still one degree of freedom left. Thus, the third step is to adjust the orientation of the radius-ulna joint. This motion is called pronation and supination of hand. In this step, the total score is calculated first. If the score is under 90, proper verbal feedback is given according to the immediate posture of user. The required rotation axis and magnitude are calculated by using the forearm orientation of the user, q_{UFArm}, and the teacher, q_{TFArm}, as in the following equation:

$$
\begin{aligned}
q_h &= q_{UFarm} \cdot (q_{TFArm})^{-1} = (w, \vec{v}_h) \\
\theta_h &= 2\sin^{-1}(w)
\end{aligned}
\tag{2}
$$

where v_h is vector of rotation axis, and θ_h is rotation angle. Positive v_h means its direction is from elbow to wrist, and negative means the opposite direction. Table 5-4 shows the words of instruction in all kinds of conditions for left radius-ulna joint. The supination and pronation are represented by "rotate outside" and "rotate inside", respectively.

After passing the three adjustment steps, the procedure finally jumps back to compare the whole upper extremity. If the total score is under 90, it starts over to check the three adjusting steps. When the total score reaches 90, the verbal instruction is "Correct." If the total score reaches 95, the verbal instruction is "Excellent."

Table 5-4 Verbal instructions under different conditions at the third adjusting step.

Direction of v_h	Sign of Θ_h	Words of instruction
+	−	Rotate hand outside
−	+	
+	+	Rotate hand inside
−	−	

5.4 System implementation

The MoRep system includes an arm suit for motion capture and an application that includes an intuitive human-computer interface, an algorithm of posture comparison, and visual and verbal feedback. This section presents the hardware, kinematic model, calibration and comparison method.

5.4.1 Hardware and system design

The hardware components of the system include the inertial measurement unit (IMU) for measuring the arm motion (i.e., the arm suit) and a computer for the interface and feedback.

The arm suit is composed of two wireless IMUs (InertiaCube, InterSense, Inc.) and one receiver, as seen in Figure 5-7(a). The sensor communicates with the receiver via 2.4GHz frequency, which itself is connected to the computer via USB. Each IMU consists accelerometer, gyroscope, magnetic sensor and wireless module. The 3D orientation of the IMU could be output in quaternion and Euler angle. According to kinesiology [46], the two IMUs are attached to the bony end of arm and forearm respectively to reduce the effect of soft tissue motion as seen in Figure 5-7(b). IMU1 captures the motion of the shoulder joint. The position of IMU2 makes it capable of simultaneously capturing the motion of the elbow joint and radius-ulna joint.

The human-computer interface is built on Microsoft .NET 3.5 SP1 Framework with Window Presentation Form (WPF). WPF utilizes DirectX for rendering the 3D simulation environment. Its advanced data binding mechanism enables displaying angles of each joint with processed values from the sensors. The program is developed with C#.

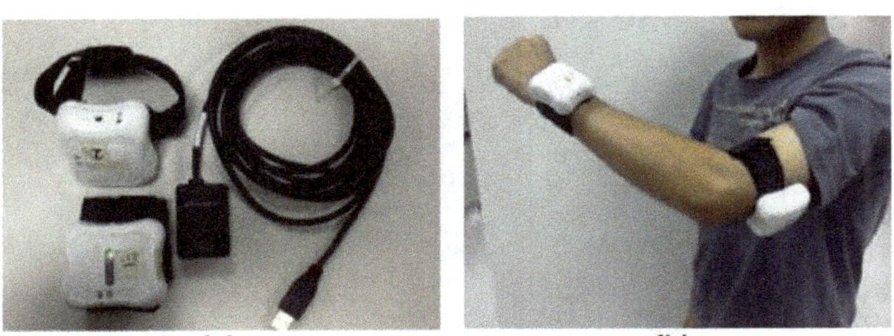

(a) (b)

Figure 5-7 (a) The arm suit and (b) the arrangement of IMU.

The algorithms of calibration, posture comparison and feedback are all integrated into this program.

5.4.2 Kinematic model

For an IMU-based MoCap system, the calculation of orientation and position are based on a kinematic model. Figure 5-8 shows the simplified kinematic model of the left upper extremity with relevant Cartesian coordinate systems (frame). This model includes two linkages (arm and forearm) and three joints (shoulder, elbow and radius-ulna). The frames of shoulder, elbow cum radius-ulna (ER) and the body are imaginary in the real world, and there are corresponding frames in the virtual world. Because the frames of the shoulder and IMU1 are attached to the same segment, the relative displacement and orientation between them are fixed. The relation between IMU2 and ER is the same. The body frame indicates the direction the user is facing. The real global frame is determined by terrestrial magnetism. We could assume that the origins of the real global, body and shoulder frame coincide.

In the following calculation, a 3 × 3 rotation matrix represents the orientation of the frame. The orientations of shoulder and ER frame with respect to real global frame could be calculated according to the following equations:

$$
\begin{aligned}
{}^{RG}R_{Shoulder} &= {}^{RG}R_{Body} \cdot {}^{Body}R_{Shoulder} \\
{}^{RG}R_{ER} &= {}^{RG}R_{Shoulder} \cdot {}^{shoulder}R_{ER}
\end{aligned} \tag{3}
$$

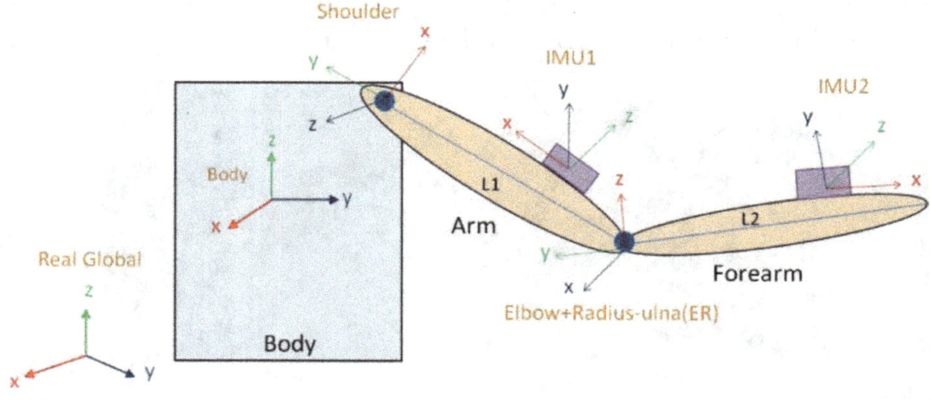

Figure 5-8 The kinematic model of left upper extremity and relevant coordinate systems in real world.

The shoulder position is defined as an origin, and the positions of elbow and wrist could be calculated by the following equations:

$$P_{shoulder} = \begin{bmatrix} 0 & 0 & 0 \end{bmatrix}$$

$$P_{Elbow} = {}^{RG}R_{Shoulder} \cdot V_1 \tag{4}$$

$$P_{Wrist} = {}^{RG}R_{ER} \cdot V_2 + P_{Elbow}$$

V_1 is a 3×1 vector, which is the position of elbow with respect to the shoulder frame. V_2 is the position of wrist with respect to the ER frame.

5.4.3 Calibration procedure

The purpose of calibration is to make the orientations of IMU1 and IMU2 in the real world correspond to the arm and forearm in virtual world. There are four steps: (1) calculate the relative orientation between IMU1 and shoulder frame, and between IMU2 and ER frame, (2) modify the facing direction, (3) transfer orientations from real world to virtual world, and (4) calculate the proportion of arm to forearm. After the calibration, the user's motion is correctly connected with his/her avatar in the virtual world. A portable height measuring device is used for aiding the calibration.

At the first step, the user sits towards the display and puts his/her left arm along the body, as seen in Figure 5-9(a). The z-axis of body frame is defined parallel to that of the real global frame, and the orientations of the shoulder and the ER frame are defined the same as the orientation of body frame at this posture. The relative orientation between IMU1 and the shoulder frame is calculated at this moment, as well as the orientation between IMU2 and the ER frame, by Equation (5).

$$^{Body}R_{Shoulder} = {}^{Body}R_{ER} = I$$

$$^{IMU1}R_{Shoulder} = (R_{P1\_IMU1})^{-1} \cdot {}^{Body}R_{Shoulder} \tag{5}$$

$$^{IMU2}R_{ER} = (R_{P1\_IMU2})^{-1} \cdot {}^{Body}R_{ER}$$

R_{P1_IMU1} means orientation of IMU1 frame at the first calibration posture.

At the second step, the user raises his/her left arm until it is parallel to the ground, and the palm points towards the floor (Figure 5-9(b)).

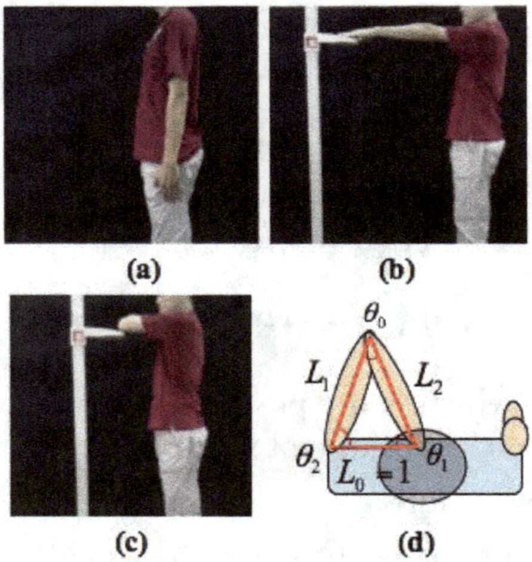

Figure 5-9 (a)(b)(c) Three postures for calibration. (d) The top-view diagram of the third calibration posture.

Assume that the *z*-axis of IMU2 frame is parallel to that of the real global frame at this posture. IMU2 frame could coincide with the real global frame by rotating about the *z*-axis. The rotation angle is calculated by the following equation:

$$R_{P2\_IMU2} = \begin{bmatrix} r_{11} & r_{12} & r_{13} \\ r_{21} & r_{22} & r_{23} \\ r_{31} & r_{32} & r_{33} \end{bmatrix} \Rightarrow \theta_M = \tan^{-1}\left(\frac{r_{21}}{r_{11}}\right) \tag{6}$$

R_{P2_IMU2} means the orientation of IMU2 frame in the second calibration posture. The range of θ_M is from −180° to 180°. The body frame could coincide with real global frame by rotating about the *y*-axis for θ_M. Thus, the relative orientation between them is determined by the following equation:

$$^{RG}R_{Body} = \begin{bmatrix} \cos\theta_M & -\sin\theta_M & 0 \\ \sin\theta_M & \cos\theta_M & 0 \\ 0 & 0 & 1 \end{bmatrix} \tag{7}$$

In the third step, the relative orientation between real and virtual global frame, $^{VG}R_{RG}$, is defined according to the orientation of the avatar in virtual reality. In our system, the relative orientation is defined as an identity matrix. Finally, the orientations of shoulder and ER frame with respect to virtual global frame could be calculated by the following equations:

$$^{VG}R_{Shoulder} = {}^{VG}R_{RG} \cdot ({}^{RG}R_{Body})^{-1} \cdot {}^{RG}R_{IMU1} \cdot {}^{IMU1}R_{Shoulder} \cdot {}^{RG}R_{Body}$$

$$^{VG}R_{ER} = {}^{VG}R_{RG} \cdot ({}^{RG}R_{Body})^{-1} \cdot {}^{RG}R_{IMU2} \cdot {}^{IMU2}R_{ER} \cdot {}^{RG}R_{Body}$$

${}^{RG}R_{IMU1}$ and ${}^{RG}R_{IMU2}$ could be directly captured by IMU1 and IMU2 respectively.

In the fourth step, the user bends his elbow and puts his wrist on the throat. The shoulder, elbow and throat form a triangle and parallel the ground (Figure 5-9(c)). Figure 5-9(d) shows the diagram of the third posture. Assume that L_0 is 1, and θ_0 and θ_2 could be calculated by captured data from IMU1 and IMU2. The proportion of arm to forearm is determined by using Sine formula:

$$\frac{L_0}{\sin\theta_0} = \frac{L_1}{\sin\theta_1} = \frac{L_2}{\sin\theta_2} \Rightarrow \frac{L_1}{L_2} = \frac{\sin(180-\theta_0-\theta_2)}{\sin\theta_2} \quad (9)$$

5.4.4 Comparison method

A comparison method that could correctly calculate in real-time is required for our system. A method that compares axes of frame is adopted [44]. It determines the similarity of two postures by outputting a score ranging between 0 and 100. The total score of left upper extremity could be determined by the following equation:

$$S_{total} = \frac{S_{arm} + S_{forearm}}{2} \quad (10)$$

S_{arm} and $S_{forearm}$ represent similarity scores of arm and forearm respectively.

5.5 Experiment

The experiment is designed to evaluate whether the feedback improves the user's posture learning. The hardware of the system includes one laptop, two IMU and one receiver. It is portable and convenient to set up in various environments. In order to reduce environmental influence on performance of the subjects, the venue is a quiet laboratory. The laptop is set on a table and connects with the receiver. When conducting this experiment, the subjects sit on a chair, wear the arm suit and face the display. Thirty normal and healthy subjects, ages 21 to 32 years (mean age = 27.03), were recruited. They were equally divided into three groups for three feedback modes (Table 5-5). Group 1 included visual and verbal feedback. Group 2 only had visual feedback but without the reminding words appearing on the display, except when the subject achieves the correct posture. Group 3 did not have any feedback or text superimposition.

Subjects perform the experiment one by one. At first, we briefly introduce the subject to the system and human-computer interface. For the subjects in the first group, we also explain the meaning of the reminder words. Then the subject wears the arm suit and goes through the calibration procedures. The body of the subject is fixed once finishing the calibration. To acclimate each subject to this system, one posture is prepared for practice before starting the formal experiment. At the beginning, the subject naturally puts left arm and left forearm downwards along the body as a preparation posture. When an assigned posture shows on the display, the subject starts to imitate it. Once the subject achieves the posture, the subject returns to the preparation posture and waits for the next assigned posture. The time taken from preparation posture to achieving posture is recorded. There are 4 assigned postures in the experiment, and each posture is repeated 4 times. The assigned postures are shown randomly in order not to let subjects easily memorize them all. Each subject has to achieve 16 assigned postures in this experiment.

Table 5-5 Detailed terms of three feedback modes.

Function	Group 1	Group 2	Group 3
Different angle of view	x	x	x
Superimposition	x	x	
Score appearance	x	x	
Color changing	x	x	
Reminding words	x		
Verbal instruction	x		
Achieving notification	x	x	x
Different angle of view	x	x	x

Each subject spends about 10–20 minutes going through the whole experiment. A questionnaire is given to evaluate comfort level, intuition level and usefulness. The answers to the questions are given on a 1–7 Likert scale from "strongly disagree" to "strongly agree." There is one specific question for Group 1 and Group 2, which is to choose the useful functions for improving posture learning. Finally, there is a free space at the end of the questionnaire, which allows for open comments.

5.6 Results and discussion

5.6.1 Analysis of the questionnaire

According to the results of the questionnaires, all subjects feel comfortable wearing the arm suit (6.3 ± 0.80). The subjects in Group 1 feel the human-computer interface is intuitive (6.2 ± 1.23), and the subjects in other groups feel it is fair (5 ± 1.34). The subjects in Group 1 agree that the feedback signals are intuitive (6.2 ± 0.92), and subjects in Group 2 feel fair (5.2 ± 1.03). The subjects in Groups 1 and 2 do not feel confused by the multi-modal feedbacks (2.7 ± 1.69). The subjects in these groups do not feel fatigued throughout the whole experiment (2.8 ± 2.20 and 3.8 ± 2.04). The subjects in Group 3 feel tired (5.5 ± 0.53) because they spent more time on preparing and adjusting their posture. The subjects in the Groups 1 and 2 agree that this MoRep system is useful for posture learning (6.3 ± 0.75), and the subjects in Group 3 feel it is fair (5.0 ± 1.15).

The subjects in Group 1 choose the useful functions according to their feeling (multiple choices). The sequence according to number of votes is: (a) superimposition, (b) different angle of view, (c) color changing, (d) audio instruction, (e) reminding words and (f) score appearance. It is obvious that vision helps most when learning new postures. However, six of ten subjects think that the audio instruction is useful for improving posture learning, especially for detailed adjustment. As for the subjects in Group 2, the order according to number of votes is: (a) superimposition, (b) different angle of view, (c) color changing and (d) score appearance, similar to the choice of Group 1. Most subjects point out that this system was fun and interesting. Some subjects suggest that it is better to add one more window for showing another angle of view, because the 3D effect of the virtual environment was not good enough. Most subjects in Group 3 complain that it was hard to achieve correct posture. They think the interface provides little information.

5.6.2 Result of the performance

The time taken from preparation posture to achieving posture, called passing time, is recorded for assessing the performance of the subjects. Ten subjects performed four assigned postures for four times in each group. A mean passing time is calculated by using the ten passing times of the same posture at the same time (Equation (11)). The ten passing times are completed by 10 subjects in the same group. Moreover, the standard deviation is calculated using Equation (12):

$$T^g_{jk} = \frac{\sum_i t^i_{jk}}{10} \tag{11}$$

$$\sigma^g_{jk} = \sqrt{\frac{\sum_i (t^t_{jk} - T^g_{jk})^2}{10-1}} \tag{12}$$

where T is mean passing time and g is group identification, which ranges between 1 and 3; i means identification of subject, which ranges between 1 and 30 and is evenly separated into three groups in order; j is posture identification, which ranges between 1 and 4, and k is the number of times, which ranges between 1 and 4. The mean passing time totals 48 under different groups, postures and number of times.

Figure 5-10 shows the chart of mean passing time with standard deviation versus number of times in Posture-1 under three different groups. The results show that the subjects in Group1 take a shorter

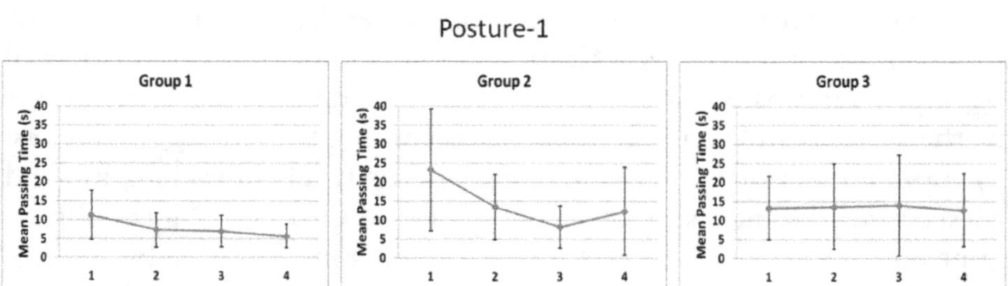

Figure 5-10 Mean passing time with standard deviation versus number of times in Posture-1 under three different group.

time to reach the posture than the subjects in other groups. The standard deviations in Group1 are also smaller than others. Thus, it proves that the combination of visual and verbal feedback helps the subjects to learn better. In Group1, the mean passing time and standard deviation consistently decrease when the number of times increases. However, There seem to be no improvement in Group 3. It shows that the feedback information also helps the subjects to memorize posture and improve performance.

The results of Posture-2 under three different groups are shown in Figure 5-11. Most subjects feel that Posture-2 is more difficult than others, and they surely spent much time adjusting their posture. The results for Group 1 are similar to Group 1 of Position-1 (Figure 5-10). Though the passing time in Group 1 is not always the shortest in each number of times, the sum of the passing times in Group 1 is smaller than those in other groups. The performances of the subjects in Group 1 are more stable and show consistent progress. However, the performances of the subjects in Groups 2 and 3 are unstable and uneven. It proves again that the combination of visual and verbal feedback is feasible and helpful, especially for difficult postures.

Figures 5-12 and 5-13 show the results of Posture-3 and Posture-4, respectively, under three different groups. Most subjects feel that Posture-3 and Posture-4 are easier to perform. The results show that the performances of the subjects in Groups 1 and 2 are similar. It shows that if a posture is easy to perform, visual feedback is enough to help the user learn a new posture. Thus, the advantage of verbal feedback is not obvious. However, There seem to be no improvement for the subjects

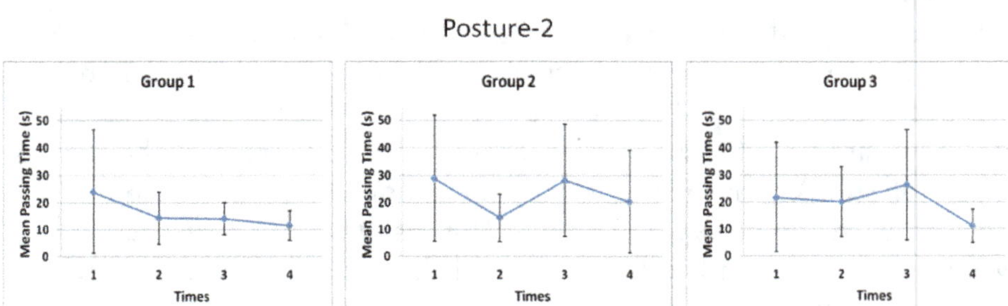

Figure 5-11 Mean passing time with standard deviation versus number of times in Posture-2 under three different groups.

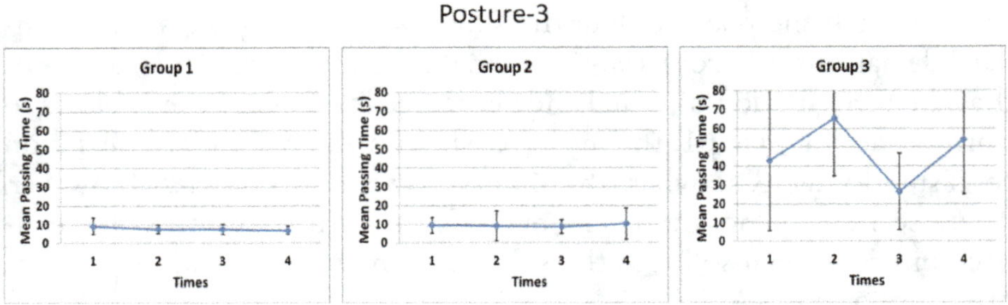

Figure 5-12 Mean passing time with standard deviation versus number of times in Posture-3 under three different groups.

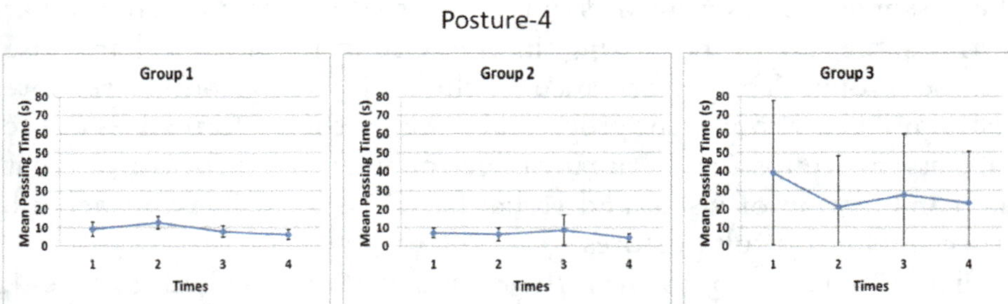

Figure 5-13 Mean passing time with standard deviation versus number of times in Posture-4 under three different groups.

in Group 3, according to their time-consuming and varying standard deviation.

In summary, all the mean passing times and standard deviations of Group 1 in the four postures decrease as the number of times increases. The subjects in Group 1 spent a shorter time to reach the correct posture compared to the subjects in other groups. Moreover, the performance of subjects in Group 1 are much more even, and become even more so with the increase of times. The results prove that the MoRep system with visual and verbal feedback is helpful to improve the performance of posture learning.

Despite the effectiveness of the system, it could still be improved. The visual feedback and the superimposition of the target arm posture on the current arm posture could suffer from occlusion. Changing the point of view of the feedback could minimize this issue. The presence of multiple feedbacks may also slow down the posture correction process, a process attributable to cognitive overload. Proper design of the feedback sequence or timing may lessen the cognitive overload of the user. All of these shortcomings are part of the future work for this project.

5.7 Conclusion

The MoRep system will be an important part of human-machine interface. A MoRep system that combines visual and verbal feedback for posture guidance of upper extremity is presented in this chapter. This system is portable and easy to set up in various environments. With the robust comparison method, and integrated visual and verbal feedback algorithm, the system can capture a user's posture, compare the posture to a reference in a database, and provide feedback to remind the user how to modify his/her posture in real time. With the intuitive human-computer interface, a user could immerse in the 3D simulation environment and seamlessly learn new postures from the virtual teacher or robot system. User tests have proved that the system with visual and verbal feedback helps the subjects to learn better and faster.

5.8 References

[1] R. Siegwart, I. R. Nourbakhsh, *Introduction to Autonomous Mobile Robots*, MIT Press, 2004.

[2] M. A. Goodrich, A. C. Schultz, "Human-robot interaction: A survey", *Foundations and Trends in Human-Computer Interaction*, vol. 1, issue 3, pp. 203–275, Feb. 2007.

[3] A. Schiele, F. C. T. van der Helm, "Kinematic design to improve ergonomics in human machine interaction", *IEEE Trans. Neural Systems and Rehabilitation Engineering*, vol. 14, issue 4, pp. 456–469, Dec. 2006.

[4] J. Y. C. Chen, E. C. Haas, M. J. Barnes, "Human performance issues and user interface design for teleoperated robots", *IEEE Trans. Systems, Man, and Cybernetics, Part C: Applications and Reviews*, vol. 37, issue 6, pp. 1231–1245, Nov. 2007.

[5] S. Collins, A. Ruina, R. Tedrake, M. Wisse, "Efficient bipedal robots based on passive-dynamic walkers", *Science*, vol. 307, no. 5712, pp. 1082–1085, Feb. 2005.

[6] J. Bruce, M. Veloso, "Real-time randomized path planning for robot navigation", in *Proc. IEEE Int. Conf. Intelligent Robots and Systems*, Lausanne, Swiss, 2002, vol. 3, pp. 2383–2388.

[7] M. A. Garcia, O. Montiel, O. Castillo, R. Sepúlveda, P. Melin, "Path planning for autonomous mobile robot navigation with ant colony optimization and fuzzy cost function evaluation", *Applied Soft Computing*, vol. 3, issue 9, pp. 1102–1110, Jun. 2009.

[8] C. Fulgenzi, A. Spalanzani, C. Laugier, "Dynamic obstacle avoidance in uncertain environment combining PVOs and occupancy grid", in *Proc. IEEE Int. Conf. Robotics and Automation*, Roma, Italy, 2007, pp. 1610–1616.

[9] B. Laurel, S. J. Mountford, The Art of Human-Computer Interface Design, Addison-Wesley Longman Publishing, 1990.

[10] S. M. M. De Rossi, N. Vitiello, T. Lenzi, R. Ronsse, B. Koopman, A. Persichetti, F. Vecchi, A. Jan Ijspeert, H. van der Kooij, M. C. Carrozza, "Sensing Pressure Distribution on a Lower-Limb Exoskeleton Physical Human-Machine Interface", *Sensors*, pp. 207–227, Nov. 2011.

[11] A. Ferreira, R. L. Silva, W. C. Celeste, T. F. Bastos Filho, M. Sarcinelli Filho, "Human–machine interface based on muscular and brain signals applied to a robotic wheelchair", *J. Phys.: Conf.* Ser.90, issue 1, 2007.

[12] D. Kragic, P. Marayong, M. Li, A. M. Okamura, "Human-machine collaborative systems for microsurgical applications", *The International Journal of Robotics*, vol. 24, no. 9, pp. 731–741, Sep. 2005.

[13] J. Scholtz, J. Young, J. L. Drury, H. A. Yanco, "Evaluation of human-robot interaction awareness in search and rescue", in *Proc. IEEE Int. Conf. Robotics and Automation*, Barcelona, Spain, 2004, vol. 3, pp. 2327–2332.

[14] de Vinci surgical system, Intuitive Surgical, Inc., http://www.intuitivesurgical.com/

[15] J. J. Abbott, A. M. Okamura, "Stable forbidden-region virtual fixtures for bilateral telemanipulation", *J. Dyn. Syst. Meas. Control*, vol. 128, no. 1, pp. 53–64, 2006.
[16] M. Kitagawa, D. Dokko, A. M. Okamura, D. D. Yuh, "Effect of sensory substitution on suture manipulation forces for robotic surgical systems", *J. Thorac. Cardiovasc. Surg*, vol. 129, no. 1, pp. 151–158, Jan. 2005.
[17] V. Hayward, O. R. Astley, M. Cruz-Hernandez, D. Grant, G. Robles-De-La-Torre, "Haptic interfaces and devices", *Sensor Review*, vol. 24, Issue 1, pp. 16–29, 2004.
[18] T. H. Massie, J. K. Salisbury, "The PHANTOM haptic interface: A device for probing virtual objects", in *Proc. Symp. Haptic Interfaces for Virtual Environment and Teleoperator Systems*, Chicago, IL, Nov. 1994.
[19] A. Mazzoli, M. Germani, and R. Raffaeli, "Direct fabrication through electron beam melting technology of custom cranial implants designed in a PHANToM-based haptic environment", *Materials & Design*, vol. 30, issue 8, pp. 3186–3192, Sep. 2009.
[20] L. Xia, C. C. Chen, J. K. Aggarwal, "Human detection using depth information by Kinect", in *Proc. IEEE Computer Society Conf. Computer Vision and Pattern Recognition Workshops*, Colorado Springs, CO, 2011, pp. 15–22.
[21] N. Miller, O. C. Jenkins, M. Kallmann, "Motion capture from inertial sensing for untethered humanoid teleoperation", in Proc. *Int. IEEE-RAS 4th Conf. Humanoid Robots*, Santa Monica, CA, 2004, vol.2, pp. 547–565.
[22] Z. Luo, C. K. Lim, W. Yang, K. Y. Tee, C. Gu, K. Doang Nguen, I-M. Chen, S. H. Yeo, "An interactive therapy system for arm and hand rehabilitation", in *Proc. Int. IEEE Conf. Robotics Automation and Mechatronics*, Singapore, 2009, pp. 9–14.
[23] R. Y. Wang, J. Popović, "Real-time hand-tracking with a color glove", *AMC Trans. Graphics*, vol. 28, issue 3, Article No. 63, Aug. 2009.

[24] K. Miura, M. Morisawa, S. Nakaoka, F. Kanehiro, K. Harada, K. Kaneko, S. Kajita, "Robot motion remix based on motion capture data towards human-like locomotion of humanoid robots", in *Proc. Int. IEEE-RAS 9th Conf. Humanoid Robots*, Paris, France, 2009, pp. 596–603.

[25] C. L. Fernando, M. Furukawa, T. Kurogi, K. Hirota, K. Sato, K. Minamizawa, S. Tachi, "TELESAR V: TELExistence surrogate anthropomorphic robot", in *Proc. ACM SIGGRAPH*, Los Angeles, CA, 2012, Article No. 23.

[26] R. O. Ambrose, H. Aldridge, R. S. Askew, R. R. Burridge, W. Bluethmann, M. Diftler, C. Lovchik, D. Magruder, F. Rehnmark, "Robonaut: NASA's space humanoid", *Intelligent Systems and their Applications*, vol. 15, issue 4, pp. 57–63, Jul. 2000.

[27] M. A. Diftler, J. S. Mehling, M. E. Abdallah, N. A. Radford, L. B. Bridgwater, A. M. Sanders, R. S. Askew, D. M. Linn, J. D. Yamokoski, F. A. Permenter, B. K. Hargrave, R. Piatt, R. T. Savely, R. O. Ambrose, "Robonaut 2 - The first humanoid robot in space", in *Proc. Int. IEEE Conf. Robotics Automation*, Shanghai, China, 2011, pp. 2178–2183.

[28] J. R. Millan, F. Renkens, J. Mouriño, W. Gerstner, "Noninvasive brain-actuated control of a mobile robot by human EEG", *IEEE Trans. Biomedical Engineering*, vol. 51, Issue 6, pp. 1026–1033, Jun. 2004.

[29] Y. H. Yin, Y. J. Fan, L. D. Xu, "EMG and EPP-integrated human-machine interface between the paralyzed and rehabilitation exoskeleton", *IEEE Trans. Inf Technol Biomed.*, vol. 16, issue 4, pp. 542–549, Jan. 2012.

[30] C. J. Bell, P. Shenoy, R. Chalodhorn, R. P. N. Rao, "Control of a humanoid robot by a noninvasive brain–computer interface in humans", *Journal of Neural Engineering*, vol. 5, no. 2, pp. 214–220, May. 2008.

[31] K. Tanaka, K. Matsunaga, "Electroencephalogram-Based Control of an Electric Wheelchair", *IEEE Trans. Robotics*, vol. 21, issue 4, pp. 762–766, Aug. 2005.

[32] M. Velliste, S. Perel, M. C. Spalding, A. S. Whitford, A. B. Schwartz, "Cortical control of a prosthetic arm for self-feeding", *Nature*, vol. 453, pp. 1098–1101, Jun. 2008.
[33] L. R. Rabiner, "A tutorial on hidden Markov models and selected applications in speech recognition", in *Proc. IEEE*, vol. 77, issue 2, pp. 257–286, 1989.
[34] O. Rogalla, M. Ehrenmann, R. Zollner, R. Becher, R. Dillmann, "Using gesture and speech control for commanding a robot assistant", in *Proc. IEEE 11th Int. Workshop on Robot and Human Interactive*, Berlin, Germany, 2002, pp.454–459.
[35] R. Dillmann, "Teaching and learning of robot tasks via observation of human performance", *Robotics and Autonomous Systems*, vol. 47, issues 2–3, pp. 109–116, Jun. 2004.
[36] J. C. P. Chan, H. Leung, J. K. T. Tang, T. Komura, "A virtual reality dance training system using motion capture technology", *IEEE Trans. Learning Technologies*, vol. 4, issue 2, pp. 187–195, Aug. 2010.
[37] S. Tsuruta, Y. Kawauchi, W. Choi, K. Hachimura, "Real-time recognition of body motion for virtual dance collaboration system," in *Proc. 17th Int. Conf. Artificial Reality and Telexistence (ICAT '07)*, Esbjerg, Denmark, 2007, pp. 23–30.
[38] K. Hachimura, K. Takashina, M. Yoshimura, "Analysis and evaluation of dancing movement based on LMA," in *Proc. 14th IEEE Int. Workshop on Robot and Human Interactive Communication (ROMAN '05)*, Nashville, TN, USA, 2005, pp. 294–299.
[39] S. Baek, S. Lee, G. J. Kim, "Motion retargeting and evaluation for VR-based training of free motions," *The Visual Computer*, vol. 19, pp. 222–242, Jul. 2003.
[40] G. Qian, F. Guo, T. Ingalls, L. Olson, J. James, T. Rikakis, "A gesture-driven multimodal interactive dance system," in *Proc. IEEE Int. Conf. Multimedia and Expo (ICME '04)*, Taipei, Taiwan, 2004, vol. 3, pp. 1579–1582.
[41] O. Portillo-Rodriguez, O. O. Sandoval-Gonzalez, E. Ruffaldi, R. Leonardi, C. A. Avizzano, M. Bergamasco, "Real-time

gesture recognition, evaluation and feed-forward correction of a multimodal Tai-Chi platform," in *Proc. 3rd Int. Workshop on Haptic and Audio Interaction Design (HAID '08)*, Jyvaskyla, Finland, 2008, vol. 5270, pp. 30–39.

[42] J. Lieberman, C. Breazeal, "TIKL: Development of a wearable vibrotactile feedback suit for improved human motor learning," *IEEE Trans. Robotics*, vol. 23, no. 5, pp. 919–926, Oct. 2007.

[43] J. Chan, H. Leung, J. Tang, T. Komura, "A virtual reality dance training system using motion capture technology," *IEEE Trans. Learning Technologies*, vol. 4, issue 2, pp. 187–195, Feb. 2011.

[44] W. -T. Yang, Z. Luo, I. -M. Chen, S. H. Yeo, "A method for comparing human postures from motion capture data," in *Proc. 18th CISM- IFToMM Symposium on Robot Design, Dynamics, and Control (ROMANSY '10)*, Udine, Italy, 2010, vol. 524, pp. 441–448.

[45] R. Shadmehr, S. P. Wise, *The Computational Neurobiology of Reaching and Pointing: A Foundation for Motor Learning*, The MIT Press, 2004.

[46] C. A. Oatis, *Kinesiology: The Mechanics and Pathomechanics of Human Movement*, 2nd ed., Lippincott Williams and Wilkins, 2008.

6. Robot education with mobile robots

Jorge Solis, Karlstad University, Sweden, Waseda University, Japan

Abstract: Even though the robot market size is still small at this moment, applied fields of robotics are gradually spreading, from the manufacturing industry to the third industry, as one of the important components to support an aging society. However; the consistent drop in birth rate in developed countries is resulting in a reduction in the number of talented students. This represents a great challenge for universities to motive young people to study science and technology. In this chapter, a survey of basic and advanced mobile robot platforms for educational purposes, at both undergraduate and graduate levels, is given. In particular, the main educational objectives with mobile robots are summarized, and two examples of mobile robot platforms for educational purposes are described, with an example of their actual use at the university level. Finally, research challenges for designing the next generation mobile robots for educational purposes are pointed out.

6.1 Introduction

Developed countries, as leading nations in scientific research and production of innovative technological products, hold very large shares in high-technology industries. In particular, as a result of several decades of research, Japan, Sweden etc. have been actively contributing to industrial development, principally in the fields of chemicals, metals, semiconductors, robotics, machinery, industrial robotics and optics. In particular, Japan is also one of the leading nations in health care, medical research and robotics. Both Japan and Sweden are the premier designers and producers of robots in the world, having created a number of entertainment and industrial robotics through commercial production and research. Such a unique environment has been made possible thanks to the great contribution and dedication of universities and private research centers.

Such great efforts have been of vital importance to the introduction of robots in different applications, such as industry [1, 2], entertainment [3–5], welfare [6, 7], education [8–10] and more (see also

Chapters 8 and 11). On the other hand, support from government in different countries has motivated the rapid increase of companies and academic institutions related to robotics field. In fact, the International Federation of Robotics (IFR) projected that sales of all types of domestic robots could reach over 9.8 million units in the period 2011–2014, with an estimated value of US$4.3 billion [11]. In addition, the Japan Robot Association (JARA), a trade group which promotes the use of robot technology, expects the Japanese market for next-generation robots (including personal robots) to reach $14 billion by 2010 and more than $37 billion by 2025. As a consequence, robotics is expected to constitute a key long-term economic strength in future decades.

However, the consistent drop in birthrate in developed countries is resulting in a reduction in the number of talented students in engineering fields. This situation may tremendously affect the industry, as it loses competitive power in the future due to the shortage of such engineers. Moreover, the curricula of engineering universities is currently lacking in practical design experience, resulting in a shortage of opportunities for promoting the creativity of students. In recent years, the proliferation of amateur robot contests (including those for elementary school children) supported by private companies and local governments have demonstrated an effective way of motivating students to get involved in the robotics field. Such contests represent an opportunity for students to actually create an object by themselves to enhance their understanding of robot technology, cultivate their practical abilities, and strive enthusiastically to achieve higher levels of accomplishment.

Even though several robotic contests have been promoted by local governments, at the university level, there are still difficulties in introducing more effective educational methods to foster the creativity of the students. In particular, most universities have been implementing Project Based Learning (PBL) to the introductory courses of mechatronics for undergraduate students. PBL is a model that organizes learning around projects. According to the definitions found in PBL handbooks for teachers, projects are complex tasks, based on challenging problems, that involve students in design, problem-solving, and decision making. Such issues give students the opportunity to work relatively autonomously, and culminate in integrating a complete system [12, 13]. This diversity of defining features, coupled with the lack of a universally accepted model or theory of Project-Based Learning, has resulted in a great variety of research and development activities. Moreover, the influence of the cultural background of students may complicate the implementation of efficient strategies to foster their creativity while solving problems.

For this purpose, several attempts to build educational robots have been made during the past few decades [8]. Since the late 1980s, when robotics was first introduced into the classroom, mobile robotics has been used in education at all levels and for various subjects [14]. Due to the fact that the fundamentals of mobile robotics is based on different engineering and science disciplines, from mechanical and electrical engineering to computer cognitive and social science [15], different types of mobile robots have been used in education in many ways. Moreover, mobile robots are designed to perform their tasks in the presence of noise, contradictory and inconsistent sensor information, and possibly in dynamic environments [16].

The development of educational robots started in the early 1980s with the introduction of the Heathkit *Hero−1* [17]. This type of robot was designed to encourage students to learn how robots are built. However, no information on the theory or principles behind the assembly was given. More recently, the majority of robotics research has focused on mobile robotics, such as surface robotics, humanoids, aerial robots, underwater robots, etc. For this purpose, several less expensive mobile robotic platforms have been introduced for educational purposes in the market [18], e.g., BEX Robotics Design system, Engino robotics, Fischertechnik, etc. Some other well-known examples include the K-Team *Hemisson* [19], which is a low-cost educational robot designed to provide an introduction to robot programming by using reduced computational power and few sensors, and the *LEGO® Mindstorms RCX*, which is a good tool for teaching early and fast robot design by using LEGO blocks [20]. In Japan, we can also find examples, such as the *RoboDesigner* kit designed to provide a general platform to enable students to build their own robots [21].

Based on the above description, there are four main educational objectives conceived for the use of mobile robots:

- To describe the fundamentals of kinematics (modeling), sensor technology (perception), actuation systems (action) and control theory (behavior).

- To describe fundamental issues related to advanced robotics and intelligent control systems, as well as their application to a broad range of fields, such as space robotics, industrial applications, health care, etc.
- To propose a complete, systematic solution of a real-world problem, as well as to design and implement the proposed solution into an embedded control system.
- To foster creativity for designing and developing novel applications for robotics.

In this Chapter, two examples of mobile robots used for education in engineering are considered. In particular, two different commercial mobile robot platforms are described: RoboDesigner® and MiniWay®. For each of them, an example of their use for educational purposes is given. RoboDesigner® is used to introduce the basics of robot technology to undergraduates in Japan, and MiniWay® is used to introduce advanced robotics topics to graduate students in Sweden.

6.2 Mobile robot platforms for education

6.2.1 RoboDesigner®

RoboDesigner® is an educational robot kit which was developed and produced by Japan RoboTech Ltd. [21]. RoboDesigner® has been designed for use as teaching material, to foster the creativity and logical thinking of future engineers and researchers. This education kit was developed within a team project supported by the government with research cooperation between the industrial and academic sectors, with Waseda University actively participating. It has been designed to simplify the understanding of robot technology while still providing sophisticated functions for advanced users and maintaining a moderate cost. This makes RoboDesigner® suitable for use as teaching material in junior high schools, colleges and universities. The standard platform set is commercially available as RDS-X03, and it includes the basic components for developing an autonomous robot. The RDS-X03 is composed of actuators, sensors, controller board, mechanical parts and C programming language software which can be easily integrated, even if the student has no experience with robots (see Figure 6-1).

The control board, RDC-101, is composed of basic components required for controlling an autonomous robot with a multi-purpose

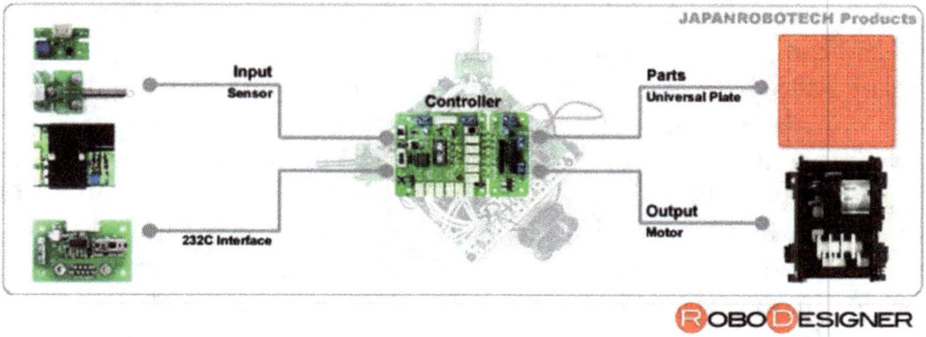

Figure 6-1 RoboDesigner® is composed of a controller, motors and sensors. Source: [10] ASME, 2009 (with permission).

electronic circuit, depending on the design requirements. The RDC-101 has five basic modules which are (see Table 6-1): peripheral interface controller, memory, motor driver, input/output port and communication port. The peripheral interface controller is based on an 8-bit CISC microcontroller. A 64 Kbits I2C memory is included. The motor driver section is located on top of the control board and it is composed of full-bridge drivers to control the forward and reverse rotational motion of the motors. The PWM control signals are linked with condition display LEDs to better understand the status of the controller. The I/O port enables the designer to connect up to five I/O devices. The rest of the connectors are used for control of the motors. Finally, in order

Table 6-1 RDC-101: hardware specifications.

Part Name	Specifications
CPU	8-bit microcontroller
Memory	64 Kbits I2C
Motor drivers	2
I/O ports	11 (Analog/digital × 3; digital × 8)
Power supply	Controller circuit: 6 ~ 12 volts Motor driver circuit: 4.5 ~ 27 volts
Communication port	RS-232C

to transmit information between the controller board and a host PC, a RS232 communication port is included.

The RDS-X03 set contains a couple of DC motors with a reduction gear box made of ABS plastic. To adjust the torque of the motor, eight different configurations can be easily selected by adjusting the position of gears, where the reduction rate can be from 1/8 up to 1/2930. This kind of flexibility enables students to learn about the relationship between gear ratio and torque. Furthermore, each motor can consume up to 2 A. In addition, it includes a touch sensor and an analog infrared sensor. The touch sensor is a simple on/off switch that uses a spring. Using this sensor, the student can easily understand the principle of a switch, as well as understanding how to process digital inputs. On the other hand, the analog infrared sensor uses a photo transistor to perceive light intensity, and it helps students understand how to process analog signals. Of course, due to the flexibility of its architecture, other types of home-made sensors can be also used.

Depending on the level of programming skills, the programming environment provides two solutions for Windows XP/2000/NT: TiColla and TiColla-CDE. TiColla provides a friendly environment where the program is created with a flowchart by a simple drag/drop method, which allows the placing of tiles into the workspace section to construct a complete program (see Figure 6-2(a)). By contrast, TiColla-CDE uses C language for writing the program (see Figure 6-2(b)).

As an example of Robodesigner® used as a mobile robot in education, an overview of the bachelor's course of Mechatronics Laboratory

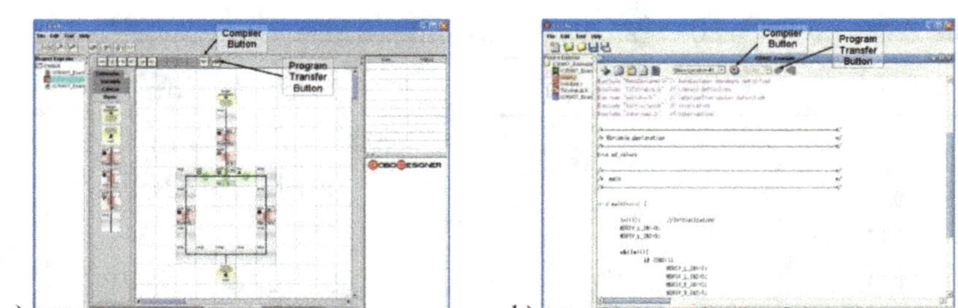

Figure 6-2 a) TiColla main window; b) TiColla-CDE programming environment. Source: [10] ASME, 2009 (with permission).

2 from the Department of Modern Mechanical Engineering at Waseda University is given [10]. In particular, this course content has been designed to cover the basic required knowledge, from microcontroller programming up to the principles of control engineering. For this course, students are required to perform different kinds of basic task experiments using the RDS-X03. The main objective of this course is to introduce to first year students the principles of mechatronics systems. At the end of the course, students are required to construct a Line-Follow Robot (LFR) based on a microcontroller (see Figure 6-3), as a part of a robot contest. The hardware specifications of the LFR are shown in Table 6-2. Regarding the TBL, each team is composed of three members. The laboratory content is organized as follows:

- LAB1 – Printed-Board Circuit Construction: Construction of a printed-board circuit and confirmation of the sensor signal processing function.
- LAB2 – Microcontroller Programming (I): Control of robot motion based on tile-programming, introduction to Tile-Programming by processing the input/output signals using RDS-X03 kit, introduction to C-Programming by processing the input/output signals, and processing of sensor information.
- LAB3 – Microcontroller Programming (II): Understanding the processing of digital sensors by controlling the characteristics of an LED, understanding the implementation prototype

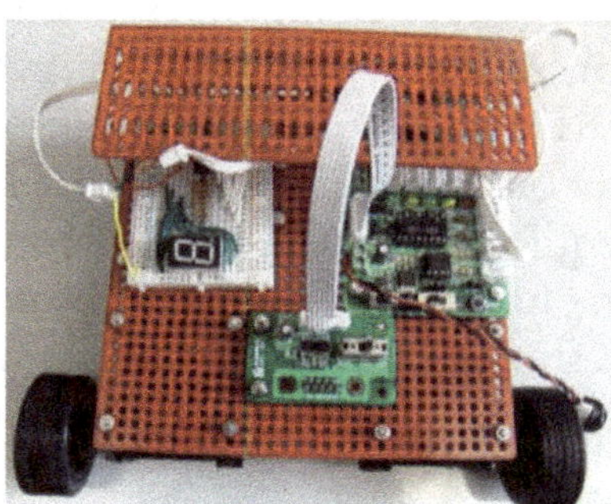

Figure 6-3 Picture of the LFR's prototype. Source: [10] ASME, 2009 (with permission).

Table 6-2 Line following robot: hardware specifications.

Part Name	Quantity
Gear Box (RDO-501)	2
IR Sensor (RDI-201)	2
Controller (RDC-101)	1
Communication Board (RDI-301)	1
Universal Plate (RDP-801)	2
Universal Pillar (RDP-803)	2
Universal Caster (RDP-806)	2
Cable (RDP-804)	5
7-Segment LED Circuit	1

functions and the application to control the speed of a motor by processing a digital sensor input, understanding the principle of processing an analog sensor input and its application to the control of a motor.

- LAB4 – Control Programming of Peripheral Devices: Control of a peripheral device, a 7-segement LED using the RDS-X03 kit, understanding the use of interruptions by controlling the functions of a 7-Segment LED, and the development of a chronometer. It is worth mentioning that the 7-segment LED is not included on the RDS-X03 set, so students are requested to construct the required circuit on a bread board.
- LAB5 – Control Programming of Systems: Understanding the principles of control theory by using an inverted pendulum system, using the RDS-X03 kit. For this purpose, an inverted pendulum system mounted on a mobile base was designed, where students are required to tune the PID control parameters (see Figure 6-4). The hardware specifications of the inverted pendulum system are shown in Table 6-3. It is worth mentioning that the analog filter was constructed by students as a required task in the LAB1.
- LAB6 – Robot Contest: The modules of the Line Following Robot are integrated, the program control is refined and the

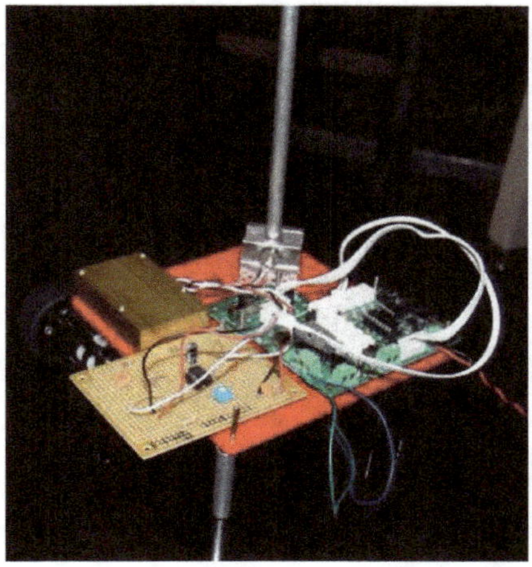

Figure 6-4 Inverted pendulum system designed for introducing the basic principles of control. Source: [10] ASME, 2009 (with permission).

Table 6-3 Inverted pendulum system: hardware specifications.

Part Name	Quantity
Controller (RDC-101)	1
Gear-Box (RDO-501)	2
Communication Board (RDI-301)	1
Universal Plate (RDP-801)	2
Universal Caster (RDP-806)	2
Pole's Mounting	2
Rotary Sensor (RDC-506)	1
Filter Circuit	1
Pole (1 m)	1

contest is held. The objective of the contest is to develop a line-following robot using an RDS-X03 kit. Each team must build one LF-Robot (see Figure 6-5), and the contest consists of building an LF-Robot which can move from START to FINISH as fast as possible (see Figure 6-6).

Figure 6-5 Pictures of LFRs constructed by students. Source: [10] ASME, 2009 (with permission).

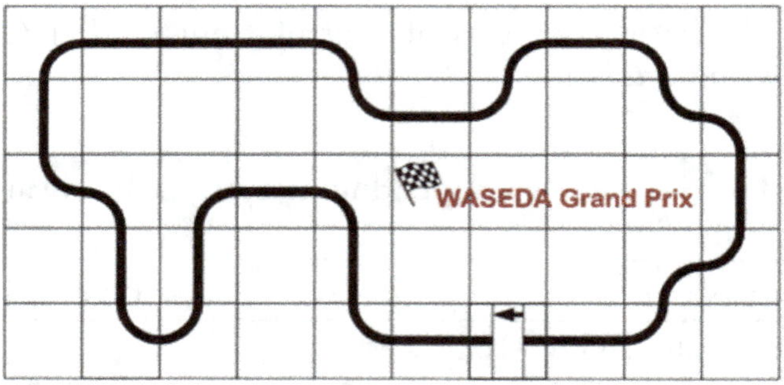

Figure 6-6 Line Course Designed for the robot contest. Source: [10] ASME, 2009 (with permission).

6.2.2 MiniWay®

A two-wheeled inverted pendulum mobile robot was designed and developed at Waseda University (see Figure 6-7). The Waseda Vehicle No. 2 Refined II was developed in order to introduce undergraduate students to the principles of robot technology [9]. Now the robot is being commercialized as MiniWay® by Japan Robotech Ltd. [21]. The MiniWay® is composed of two actuated wheels, a general-purpose control board (see Figure 6-8(a)), an adjustable weighting bar attached to the pendulum, a gyro and accelerometer sensors, a remote controller (see Figure 6-8(b)), and two optional mechanisms that can be easily attached/detached from the main body of the robot (see Table 6-4).

Figure 6-7 A two-wheeled inverted pendulum mobile robot was designed and developed at Waseda University and is now being commercialized as MiniWay® by Japan Robotech Ltd. Source: [22] 13th World Congress in Mechanism and Machine Science, 2011 (with permission).

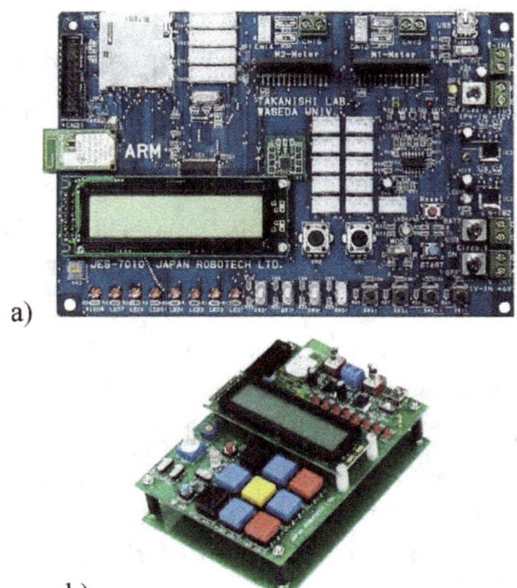

Figure 6-8 a) General purpose control board; b) Remote controller for MiniWay. Source: [23] IGI Global, USA, 2012 (with permission).

Table 6-4 Specifications of MiniWay®.

Parameter	Specifications
Height [mm]	495
Weight [kg]	3.8
DOFs	2-DOFs
Microcontroller	STM32F103VB × 1
Sensors	Accelerometer × 1 Red Gyro × 1 Optical Encoder x 2
Motor	RDO-37BE50G9 (12 volts) × 2
Power supply	Battery: 6 [V] × 1 RC–Battery: 12 [V] × 1
Remote controller	ZigBee: 2.6 GHz

In particular, the control board consists of a 32-bit ARM microcontroller, 10 general I/O ports, 2 motor drives, a LCD display, 8 LEDs, a Zigbee module and 2 servo connectors. In addition, a 3D simulator has been developed by computing the equations of motion so that students may compare the theoretical and real response of the robot (see Figure 6-9).

A simulator was developed at Waseda University [22] in order to enable students to perform simulation experiments with MiniWay® (see Figure 6-9). The purpose of the simulator is to allow the students to carry out preliminary tests of their designed controller before doing experiments with the hardware. In order to derive motion equations, the mechanical model of the robot has been defined (see Figure 6-10(a)). In this case, it is assumed that the body coordinate system is attached to the chassis of the robot (X_{ch}, Y_{ch}, and Z_{ch}). The second coordinate system is placed in the mass centre of the robot (X_c, Y_c, and Z_c). θ is defined as the angular rotation of the chassis with respect to the ground, ψ_L and ψ_R are the angular rotation of the wheels with respect to the chassis, and φ_L and φ_R are the angular rotation of the wheels with respect to the ground (see Figure 6-10(b)). The last angular relation that needs to be specified is the change in heading direction (see Figure 6-10(c)), which in our consideration is described by

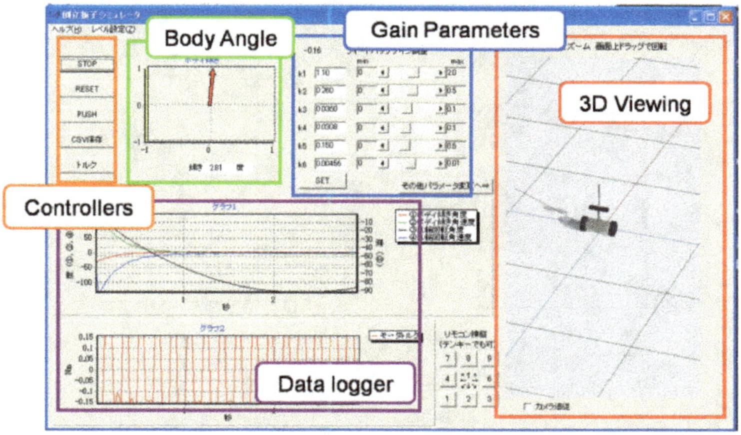

Figure 6-9 Picture of the 3D simulator developed at Waseda University and implemented for the MiniWay®. Source: [22] 13th World Congress in Mechanism and Machine Science, 2011 (with permission).

a. For the derivation of motion equations, we use the Euler-Lagrange method. Therefore, the kinetic energy of the system has been derived. In particular, the total kinetic energy, which is shown in Equation (1), has been derived at Waseda University and then implemented in the simulator for MiniWay®. The first term computes the final kinetic energy from the translational motion. The second term represents the kinetic energy from the angular motion of the robot. Finally, the last term determines the kinetic energy coming from the rotational motion of the mass centre of the chassis (where the rotational velocity vector is considered in respect to the centre of the mass coordinate frame).

$$K = \frac{1}{2}mv_c^2 + \frac{1}{2}(I_w + I_{rot})(\dot{\psi}_L^2 + \dot{\psi}_R^2) + \frac{1}{2}I_{xx}\omega_x^2 + \frac{1}{2}I_{yy}\omega_y^2 + \frac{1}{2}I_{zz}\omega_z^2 \quad (1)$$

The block diagram of the control system designed at Waseda University [22] and then implemented for the MiniWay is based on a feedback controller (see Figure 6-11). In particular, the rate gyro sensor signal measures the body angular velocity (θ') and the encoder

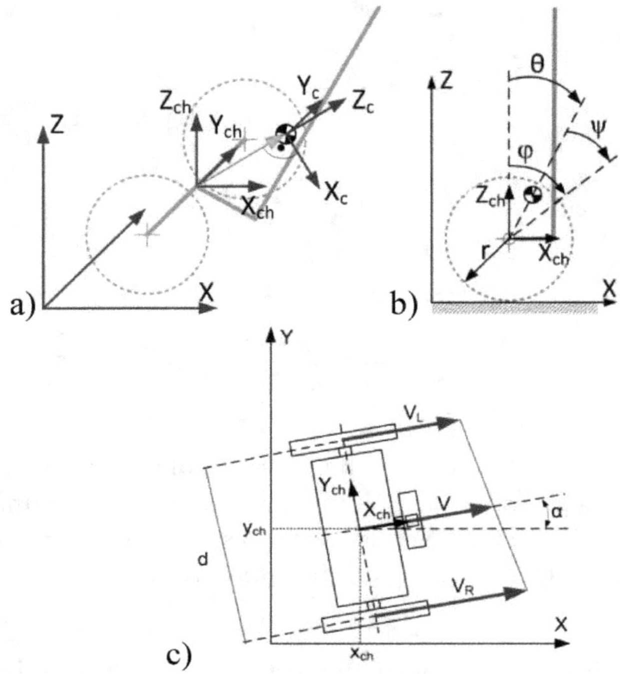

Figure 6-10 Model reference: a) Coordinate systems; b) Definition of angles; c) Translational velocity. Source: [22] 13th World Congress in Mechanism and Machine Science, 2011 (with permission).

measures the wheel rotational angle (φ). Because the drift on the signal obtained from the gyro is extremely small, the use of a high-pass filer is not required. Therefore, only a low-pass filter is used to compute the body angular velocity (θ'), where the cut-off frequency is 0.32 Hz. In order to compute the body angle, the wheel angular velocity, the body angular velocity and wheel angle are integrated and derived respectively. In order to control all the parameters, a feedback controller has been implemented by using Equation (2), where the $k_1 \sim k_6$ parameters are the gain coefficients of the controller, which are tuned to assure the stabilization of the system. Furthermore, a current feedback controller has been implemented by Equation (3), where the parameter k_7 is tuned to assure accurate control of the command current to each motor. As for the command control signal, the θ'_{REF}, φ'_{REF} and α'_{REF} are set to zero, while the other commands are sent by a remote controller.

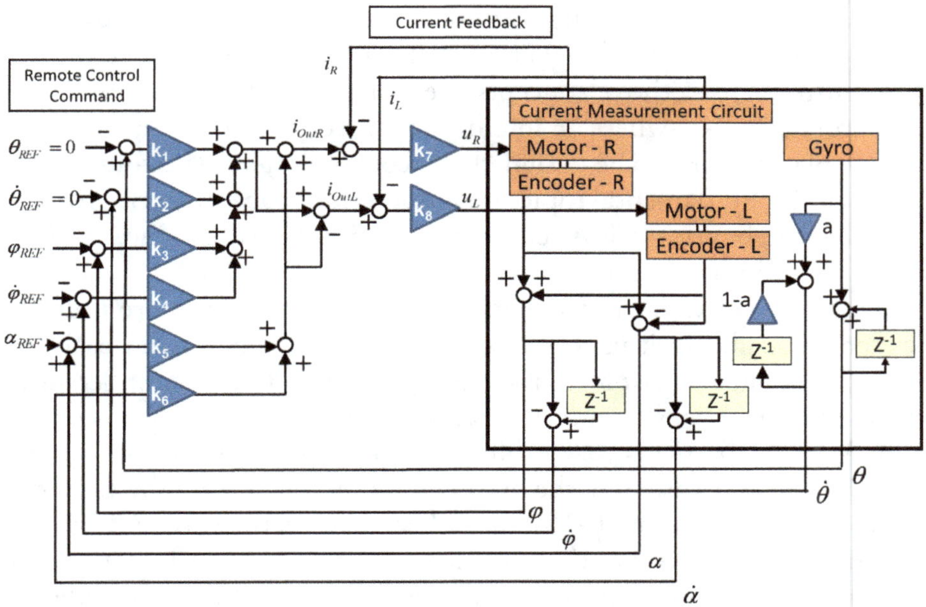

Figure 6-11 Control block diagram designed at Waseda University and implemented for MiniWay®. Source: [22] 13th World Congress in Mechanism and Machine Science, 2011 (with permission).

$$i_{outR} = k_1 \cdot \theta + k_2 \cdot \theta' + k_3 \cdot (\varphi - \varphi_{REF}) + k_4 \cdot (\varphi' - \varphi'_{REF})$$
$$+ k_5 \cdot (\alpha - \alpha_{REF}) + k_6 \cdot (\alpha' - \alpha'_{REF}) \quad (2)$$
$$u_R = k_7 \cdot (i_{outR} - i_R)$$

$$i_{outL} = k_1 \cdot \theta + k_2 \cdot \theta' + k_3 \cdot (\varphi - \varphi_{REF}) + k_4 \cdot (\varphi' - \varphi'_{REF})$$
$$- k_5 \cdot (\alpha - \alpha_{REF}) - k_6 \cdot (\alpha' - \alpha'_{REF}) \quad (3)$$
$$u_L = k_8 \cdot (i_{outL} - i_L)$$

As an example of the use of MiniWay® as a mobile robot in education, the master course on Advanced Robotics and Intelligent Control (ELAD16) from the Department of Engineering and Physics at Karlstad University is briefly described [24]. In this course, the fundamentals of robotic systems, with special focus on understanding different advanced robotic platforms and intelligent control algorithms, are covered during the lectures. Then, those concepts are illustrated through mandatory laboratory exercises where an intelligent control system on an advanced robotic platform has to be designed as a course project. The course content includes fundamental topics on autonomous robots, service robots, bio-robotics, haptics, advanced control and artificial intelligent algorithms for robotic platforms.

During the course, three obligatory laboratory exercises are utilized. Students usually work in teams (three or four members) and each exercise is designed to reinforce the basic principles developed in lectures. In the first practice, students are asked to tune the control gains of the MiniWay®. In particular, two different controller strategies are tested, where in the first case only the gyro is used for computing the tilt angle of the chassis (see Figure 6-10). In the second case, the gyro and accelerometer are used for computing the tilt angle of the chassis. In both cases, the control laws defined in Equation (2) and Equation (3) are considered.

In the second exercise, students are asked to verify the effect of changing the microprocessor clock time setting (100 ms and 10 ms), as well as understanding the effect of the ground stability (tune the parameter value of *mlt,* in msec) to have the best performance where a small value of *mlt* is preferred in a ground with high friction coefficient (i.e., a carpet). A large value of *mlt* is preferred in a ground with low friction coefficient (i.e., flooring). Finally, in the last exercise, students are asked to design and implement a fuzzy logic controller to assure the stability of the two-wheeled inverted pendulum robot. For this purpose, students are asked to define the membership functions and implement them in the MiniWay®.

As for the course project, students were required to state a real-world problem (see also Chapters 1, 4 and 5 from the book Mobile Robots for Dynamic Environments), to propose a complete, systematic solution of the problem, and to implement it in the MiniWay® by its program environment. At the end, students were asked to give a presentation and to submit a complete report for each team where the problem, the proposed methods, their implementation, and the experimental results are clearly described. As for the fall course in 2011, four different projects were proposed by the students.

One example of a course project proposed by students is related to an intelligent robot navigation based on Neural Networks. The main objective of this project was to implement intelligent control for the robot's navigation, implement advanced vision processing to identify landmarks, accomplish the task of maneuvering in the workspace avoiding obstacles and moving towards the final goal. For this purpose, the two-wheeled inverted pendulum robot was equipped with two IR sensors and a webcam (see Figure 6-12). For this project the number of inputs for the Neural Network was 6 (2 × IR sensors, Colour code, *PosX*, *PosY* and *HW*). The number of outputs for the NN was set to 1. In this project the NN output was trained to simulate the navigation control of the Miniway®.

Another example of a course project proposed by students is related to the development of a layered fuzzy behavior-based navigation for an autonomous mobile robot. The main objective was to design and construct an autonomous mobile robot capable of coming up with collision-free trajectories, in static as well as dynamic environments containing obstacles, between a start and a goal configuration. Along this route, artificial landmarks were present that give a general indication as to where the next set of landmarks could be found. Multiple artificial landmarks of varying colors were hence used having different interpretations. For this purpose; the mobile base robot was equipped with two IR sensors,

Figure 6-12 Picture of MiniWay® with IR sensors and a webcam used for the navigation control. Source: [24] 2nd IFToMM Asian Conference on Mechanism and Machine Science, 2012 (with permission).

a webcam and a laptop computer mounted on the robot chassis (see Figure 6-13(a)). Furthermore, layered control architecture for the developed autonomous mobile robot was proposed (see Figure 6-13(b)). We can see the three layers: a Supervision Layer, a Behavior Layer and a Locomotion Layer. The Supervision Layer is used for behavior coordination and fusion. The Behavior Layer is used to relate the states and actions to robot physical sensors and the different motion patterns in the locomotion layer. The Locomotion Layer receives commands to be executed directly from the behavior layer.

6.2.3 Learning outcomes

From the two examples of mobile robots for educational purpose described above, it is possible to understand that mobile robotics can facilitate the hands-on practice of the following main topics (both at the basic and advanced level):

- Kinematics, e.g., deriving equations of motion
- Sensor technology, e.g., designing and implementing digital filters, designing and implementing sensor fusion techniques
- Actuation systems, e.g., analyzing the effect of backlash
- Control theory, e.g., designing and implementing PID controllers as well as intelligent controllers

By understanding these main topics, it is then possible for students to solve real-world problems, such as autonomous navigation, line following, stabilizing an inverted pendulum, etc. In particular, the students are able to use advanced tools for designing and implementing real-time applications (e.g., Matlab/Simulink, Eclipse, etc.) into the microprocessor. In fact, it is worth mentioning that one of the students has proposed and developed an Intelligent Carrying Assistant Robot (*iCAR*) for real application in medical facilities, as part of his master thesis degree. As a result of this, a research paper was presented at the 14th Mechatronics Forum International Conference [25]. As we have noticed in the examples given from students of the course ELAD16 at Karlstad University, the use of MiniWay® motivates the students to propose a complete, systematic solution of real-world problems (e.g., navigation control in unstructured environments). The complexity of dealing with such problems promoted creative thinking.

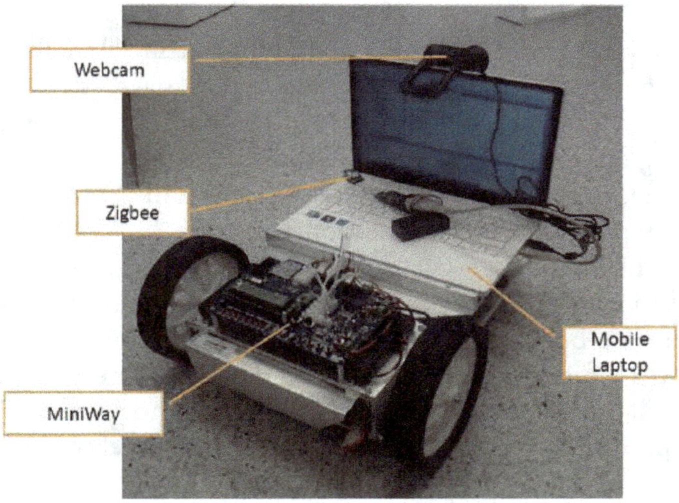

a)

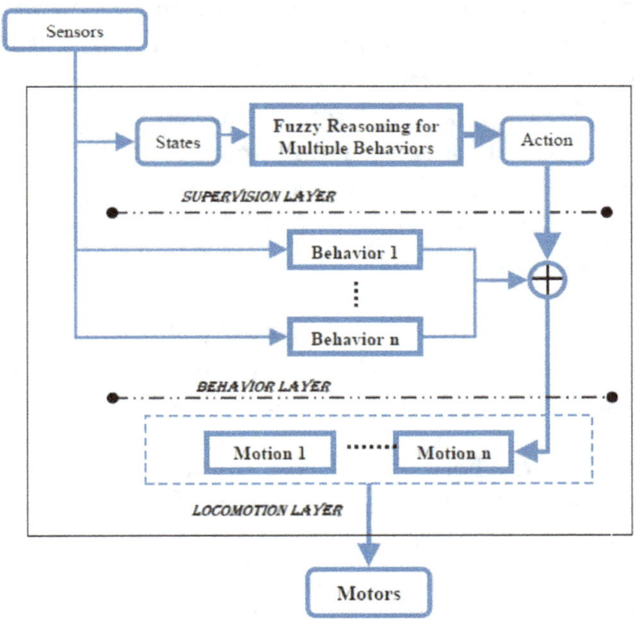

b)

Figure 6-13 a) Picture of MiniWay® with IR sensors and a webcam used for the control navigation; b) Control architecture. Source: [24] 2nd IFToMM Asian Conference on Mechanism and Machine Science, 2012 (with permission).

6.3 Research problems and trends

The recent advances in computer technology, artificial intelligence, speech synthesis and understanding, and remote control have led to breakthroughs in robotic technology. On the other hand, human/robot interaction (HRI) is an emerging topic of interest for both basic research and costumer application (see Chapter 5 from the book Designs and Prototypes for Mobile Robots). In particular, in the evolution of HRI, at least three levels of interaction can be identified [26]: cooperative robots [27], autonomous robots [28] and Haptic Interfaces [29]. Even though different robot educational platforms have been proposed [8, 21, 22, 30, 31], most of them have been developed to fulfill at least one of the levels of interaction. Therefore, there is still a research challenge in introducing novel educational robots that fulfill the following requirements at different levels of interaction:

- To introduce the fundamentals and advanced topics of human-robot interaction (e.g., human-in-the-loop control) at the different levels of interaction (e.g., cooperative, autonomous and multimodal interactive interface);
- To display the characteristics of a dynamical system but without extreme complication (i.e., humanoid robot);
- To be affordable to both university and students;
- To introduce effective ways to evaluate the progress of the students' learning.

On the other hand, from the designing point of view, there are several challenges dealing with the well-known tradeoff between the educational features and the cost affordability. Some of those challenges are:

- Enhancing the hardware design, simplifying assembly, increasing robustness, adding useful educational features, e.g., AFRON "10 Dollar Robot" design challenge [32].
- Educational features designed to be accessible to students with and without disabilities, e.g., ROBUS [33].
- Cost affordable to both students and teachers from compulsory schools, e.g., robot kit based on the UNO Arduino [34].
- Enhancing the entertainment aspect, e.g., including aspects of art with the MiniWay® to learn basic theory on signal processing and recognition systems [23].

The recent trends of mobile robots for education are focused on introducing science, technology, engineering and mathematics (STEM) education for students. As such, they are not only able to understand the principles of robot technology but also to learn other related topics of actual importance to introducing robotics to the third industry, such as renewable energy (e.g., solar panels), bi-directional remote controllers (e.g., haptic desktop from Novint Falcon® [31]), and motion-based controllers (e.g., LeapMotion), etc.

6.4 Closure

In this Chapter, a survey of mobile robots for education has been given. In particular, two platforms were described to introduce the basics of robot technology to undergraduate students, and to introduce advanced topics in robotics to graduate students. In addition, some of the possible learning outcomes gained by students from those described platforms were pointed out. From these, the use of mobile robots for educational purposes helps students to understand fundamental theory, to propose solutions by applying their knowledge and using advanced tools for designing, and to verify the effectiveness in unstructured environments. Finally, some research trends to introduce novel educational robots were introduced, and the current design challenges were summarized.

6.5 References

[1] Kaneko, K., Kanehiro, F., Kajita, S., Yokoyama, K., Akachi, T., Ota, S., Isozumi, T., 2002, Design of Prototype Humanoid Robotics Platform for HRP, *Proc. IEEE/RSJ Int. Conference on Intelligent Robots and Systems*, pp. 2431–2436.

[2] TMSUK: http://www.tmsuk.co.jp/english/robots.html (accessed 9 March 2014).

[3] Ishida, T., Kuroki, Y., Yamaguchi, J., Fujita, M., Doi, T., 2001, Motion entertainment by a small humanoid robot based on OPEN-R, *Proc. of the IEEE/RSJ International Conference on Intelligent Robots and Systems*, pp. 1079–1086.

[4] Sakagami, Y., Watanabe, R., Aoyama, C., Matsunaga, S., Higaki, H., Fujimura, K., 2002, The intelligent ASIMO: System overview and integration, *Proc. IEEE/RSJ Int. Conf. on Intelligent Robots and Systems*, pp. 2478–2483.

[5] Solis, J., Takanishi, A. et al., 2008, Understanding the mechanisms of the human motor control by imitating flute playing with the Waseda Flutist Robot WF-4RIV, *Mechanism and Machine Theory Journal*, **44**(3), pp. 527–540.

[6] Shibata, T., Mitsui, T., Wada, K., Kumasaka, T., Tagami, K., Tanie, K., 2001, Therapy of children assisted by mental commit robot, *Proc. IEEE/ASME International Conference on Advanced Intelligent Mechatronics*, p. 4.

[7] Sugahara, Y., Lim, H., Takanishi, A. et al., 2005, Walking Up and Down Stairs Carrying a Human by a Biped Locomotor with Parallel Mechanism, *Proc. of the IEEE/RSJ International Conference on Intelligent Robots and Systems*, pp. 3425–3430.

[8] Miller, D. et al., 2008, Robots for Education, *Springer Handbook of Robotics*, Siciliano, B.; Khatib, O. (Eds.), pp. 1287–1290.

[9] Solis, J., Takanishi, A. et al., 2008, Towards understanding the suture/ligature skills during the training process by using the WKS-2RII, *International Journal of Computer Assisted Radiology and Surgery*, **3**(3–4), pp. 231–239.

[10] Solis, J., Takanishi, A., 2009, Introducing Robot Technology to Undergraduate Students at Waseda University, *Proceedings of the ASME Asia-Pacific Engineering Education Congress*, 10004.

[11] International Federation of Robotics: http://www.ifr.org/service-robots/statistics/ (accessed 29 November 2013).

[12] Jones, B.F., Rasmussen, C.M., Moffitt, M.C., 1997, Real-life problem solving: A collaborative approach to interdisciplinary learning, *American Psychological Association*.

[13] Thomas, J.W., Mergendoller, J.R., Michaelson, A., 1999, *Project-Based Learning: A Handbook for Middle and High School Teachers*, The Buck Institute for Education.

[14] Malec, J., 2001, Some Thoughts on Robotics for Education, *2001 AAAI Spring Symposium on Robotics and Education*.

[15] Siegwart, R., Nourbakhsh, I., 2004, *Introduction to Autonomous Mobile Robots*, MIT Press, Cambridge, MA, USA.

[16] Ali, W.G., 2011, A semi-autonomous mobile robot for education and research, *Journal of King Saud University – Engineering Sciences*, **23**(2), pp. 131–138.

[17] Heath Company: http://www.hero-1.com/broadband/ (accessed 11 March 2014).

[18] Demetriou, G.A., 2011, Mobile Robotics in Education and Research, *Mobile Robots*, InTech Publications, pp. 27–48.

[19] K-TEAM: http://www.k-team.com/ (accessed 17 March 2014).

[20] LEGO: http://mindstorms.lego.com/ (accessed 16 March 2014).

[21] Japan Robotech: http://www.japan-robotech.com/eng/index.html/ (accessed 16 March 2014).

[22] Solis, J., Takanishi, A. et al., 2011, Development of the Waseda Wheeled Robot No. 2 Refined II and Pilot Experiments with Undergraduate Students" *13th World Congress in Mechanism and Machine Science*, A20–345.

[23] Solis, J., Takanishi, A., 2012, Human-Friendly Robots for Entertainment and Education, *Service Robots and Robotics: Design and Application*, Cecarelli, M. (Eds.), IGI Global: Hershey PA, USA, pp. 130–153.

[24] Solis, J., 2013, Introduction of Robotics to Master Students at Karlstad University, *Proceedings of the 2nd IFToMM Asian Conference in Mechanism and Machine Science*, ID83.

[25] Solis, J., Teshome, T.D., 2014, Development of a human friendly robot vehicle for carrying-medical tools: Embodying perceptual capabilities, *Proc. of the 14th Mechatronics Forum International Conference*, pp. 370–376.

[26] Solis, J., 2004, Robotic control systems for learning and teaching human skills, PhD Thesis, SSSA/PERCRO, p. 226.

[27] Yokokohji, Y. et al., 2003, Toy Problem as the Benchmark Test for Teleoperation Systems, *Advanced Robotics*, **17**(3), pp. 253–273.

[28] Atkenson, C.G., 2000, Using humanoid robots to study human behavior, *IEEE Intelligent System Application*, **15**(4), pp. 46–56.

[29] Gillespie, B., Cutkosky, M., 1996, Stable user-specific rendering of the virtual wall, *Proc. of the ASME International Mechanical Engineering Conference and Exposition,* **58**, pp. 397–406.

[30] Aldeberan: http://www.aldebaran-robotics.com/ (accessed 17 March 2014).

[31] Novint: http://www.novint.com/index.php/novintfalcon/ (accessed 15 March 2014).

[32] AFRON: http://www.robotics-africa.org/afron-design-challenges.html (accessed 16 March 2014).

[33] Michaud, F., 2007, Engineering Education and the Design of Intelligent Mobile Robots for Real Use, *Intelligent Automation & Soft Computing,* **13**(1), pp. 19–28.

[34] Junior, L.A., Neto, O.T., Hernandez, M.F., Martins, P.S., Roger, L.L., Guerra, F.A., 2013, A Low-Cost and Simple Arduino-Based Educational Robotics Kit, Multidisciplinary Journals in Science and Technology, *Journal of Selected Areas in Robotics and Control,* **3**(12), pp. 1–7.

www.ingramcontent.com/pod-product-compliance
Lightning Source LLC
LaVergne TN
LVHW011514200926
848484LV00001B/36